D1608797

PERIODIC TABLE OF THE ELEMENTS

Metals and Metalloids

Monica Halka, Ph.D., and
Brian Nordstrom, Ed.D.

METALS AND METALLOIDS

Facts On File, Inc.
An imprint of Infobase Publishing
132 West 31st Street
New York NY 10001

Library of Congress Cataloging-in-Publication Data
Halka, Monica.
Metals and metalloids / Monica Halka and Brian Nordstrom.
p. cm. — (Periodic table of the elements)
Includes bibliographical references and index.
ISBN 978-0-8160-7370-2
1. Metals. 2. Semimetals. 3. Periodic law. I. Nordstrom, Brian II. Title.
QD171.H25 2010
546'.3—dc22 2009049369

Facts On File books are available at special discounts when purchased in bulk quantities for businesses, associations, institutions, or sales promotions. Please call our Special Sales Department in New York at (212) 967-8800 or (800) 322-8755.

You can find Facts On File on the World Wide Web at http://www.factsonfile.com

Text design by Erik Lindstrom
Composition by Hermitage Publishing Services
Illustrations by Dale Williams
Photo research by Tobi Zausner, Ph.D.
Cover printed by Yurchak Printing, Landisville, Pa.
Book printed and bound by Yurchak Printing, Landisville, Pa.

Printed in the United States of America

PERIODIC TABLE OF THE ELEMENTS

Metals and Metalloids

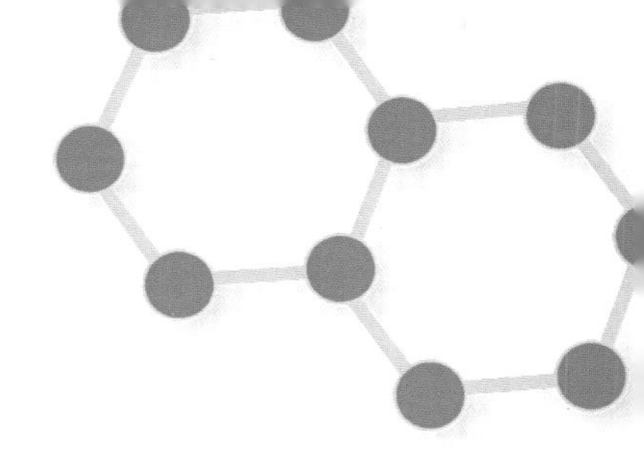

Contents

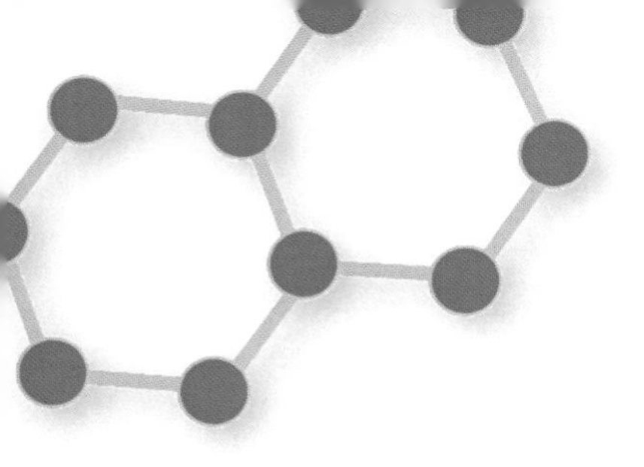

Preface

Speculations about the nature of matter date back to ancient Greek philosophers like Thales, who lived in the sixth century B.C.E., and Democritus, who lived in the fifth century B.C.E., and to whom we credit the first theory of *atoms.* It has taken two and a half millennia for natural philosophers and, more recently, for chemists and physicists to arrive at a modern understanding of the nature of *elements* and *compounds.* By the 19th century, chemists such as John Dalton of England had learned to define elements as pure substances that contain only one kind of atom. It took scientists like the British physicists Joseph John Thomson and Ernest Rutherford in the early years of the 20th century, however, to demonstrate what atoms are—entities composed of even smaller and more elementary particles called *protons, neutrons,* and *electrons.* These particles give atoms their properties and, in turn, give elements their physical and chemical properties.

After Dalton, there were several attempts throughout Western Europe to organize the known elements into a conceptual framework that would account for the similar properties that related groups of elements exhibit and for trends in properties that correlate with increases in atomic weights. The most successful *periodic table* of the elements was designed in 1869 by a Russian chemist, Dmitri Mendeleev. Mendeleev's method of organizing the elements into columns grouping elements with similar chemical and physical properties proved to be so practical that his table is still essentially the only one in use today.

While there are many excellent works written about the periodic table (which are listed in the section on further resources), recent scientific investigation has uncovered much that was previously unknown about nearly every element. The Periodic Table of the Elements, a six-volume set, is intended not only to explain how the elements were discovered and what their most prominent chemical and physical properties are, but also to inform the reader of new discoveries and uses in fields ranging from astrophysics to material science. Students, teachers, and the general public seldom have the opportunity to keep abreast of these new developments, as journal articles for the nonspecialist are hard to find. This work attempts to communicate new scientific findings simply and clearly, in language accessible to readers with little or no formal background in chemistry or physics. It should, however, also appeal to scientists who wish to update their understanding of the natural elements.

Each volume highlights a group of related elements as they appear in the periodic table. For each element, the set provides information regarding:

- the discovery and naming of the element, including its role in history, and some (though not all) of the important scientists involved;
- the basics of the element, including such properties as its atomic number, atomic mass, electronic configuration, melting and boiling temperatures, abundances (when known), and important isotopes;
- the chemistry of the element;
- new developments and dilemmas regarding current understanding; and
- past, present, and possible future uses of the element in science and technology.

Some topics, while important to many elements, do not apply to all. Though nearly all elements are known to have originated in stars or stellar explosions, little information is available for some. Some others that have been synthesized by scientists on Earth have not been observed in stellar spectra. If significant astrophysical nucleosynthesis research exists, it is presented as a separate section. The similar situation applies for geophysical research.

Special topic sections describe applications for two or more closely associated elements. Sidebars mainly refer to new developments of special interest. Further resources for the reader appear at the end of the book, with specific listings pertaining to each chapter, as well as a listing of some more general resources.

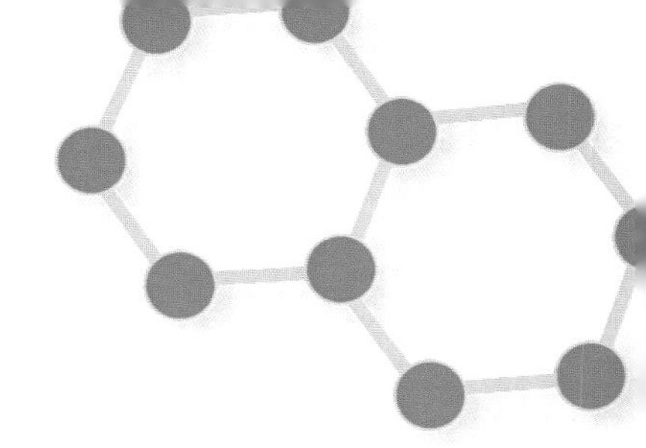

Acknowledgments

First and foremost, I thank my parents, who convinced me that I was capable of achieving any goal. In graduate school, my thesis adviser, Dr. Howard Bryant, influenced my way of thinking about science more than anyone else. Howard taught me that learning requires having the humility to doubt your understanding and that it is important for a physicist to be able to explain her work to anyone. I have always admired the ability to communicate scientific ideas to nonscientists and wish to express my appreciation for conversations with National Public Radio science correspondent Joe Palca, whose clarity of style I attempt to emulate in this work. I also thank my coworkers at Georgia Tech, Dr. Greg Nobles and Ms. Nicole Leonard, for their patience and humor as I struggled with deadlines.

—Monica Halka

In 1967, I entered the University of California at Berkeley. Several professors, including John Phillips, George Trilling, Robert Brown, Samuel Markowitz, and A. Starker Leopold, made significant and lasting impressions. I owe an especial debt of gratitude to Harold Johnston, who was my graduate research adviser in the field of atmospheric chemistry. I have known personally many of the scientists mentioned in the Periodic Table of the Elements set: For example, I studied under Neil Bartlett, Kenneth Street, Jr., and physics Nobel laureate Emilio Segrè. I especially cherish having known chemistry Nobel laureate Glenn

Seaborg. I also acknowledge my past and present colleagues at California State University; Northern Arizona University; and Embry-Riddle Aeronautical University, Prescott, Arizona, without whom my career in education would not have been as enjoyable.

—Brian Nordstrom

Both authors thank Jodie Rhodes and Frank Darmstadt for their encouragement, patience, and understanding.

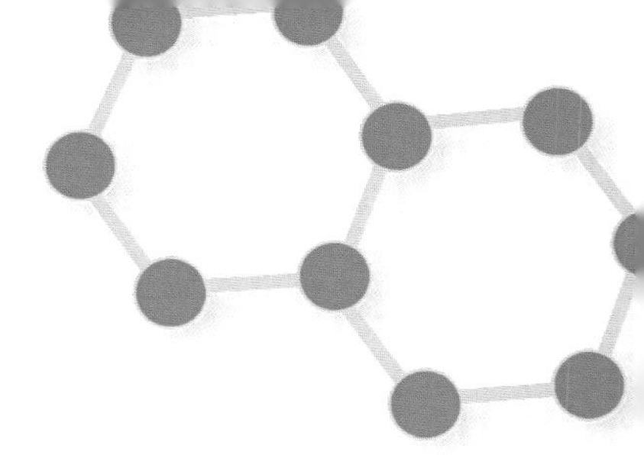

Introduction

Materials that are good conductors of electricity are generally considered to be metals. One important use of metals, in fact, is the capability to be used in electrical circuitry. All of the metallic elements on Earth exist in its crust, mantle, or core.

As an important introductory tool, the reader should note the following properties of metals in general:

1. The atoms of metals tend to be larger than those of nonmetals. Several of the properties of metals result from their atomic sizes.
2. Metals exhibit high electrical conductivities. High electrical conductivity is the most important property that distinguishes metals from nonmetals.
3. Metals have low electronegativities; in fact, they are electropositive. This means that the atoms of metals have a strong tendency to lose electrons to form positively charged ions, a tendency that is responsible for metals' electrical conductivities.
4. Metals have low electron affinities. This means that gaining additional electrons is energetically unfavorable. Metal atoms would much rather give up one or more electrons than gain electrons.
5. Under normal conditions of temperature and pressure, with the exception of mercury, all metals are solids at room

temperature. In contrast, many nonmetals are gases, one is a liquid, and only a few are solids. The fact that so many metals exist as solids means that metals generally have relatively high melting and boiling points under normal atmospheric conditions.

6. In their solid state, metals tend to be malleable and ductile. They can be shaped or hammered into sheets, and they can be drawn into wires.
7. Metals tend to be shiny, or lustrous.

While scientists categorize the chemical elements as metals, nonmetals, and metalloids largely based on the elements' abilities to conduct electricity at normal temperatures and pressures, there are other distinctions taken into account when classifying the elements in the periodic table. The post-transition metals, for example, are metals, but they have such special properties that they are given their own classification. The same is true for the metalloids. Both families of elements appear to the right of the transition metals block, which dominates the center of the periodic table.

Metals and Metalloids presents the current scientific understanding of the physics, chemistry, geology, and biology of these two families of elements—the post-transition metals and the metalloids—including how they are synthesized in the universe, when and how they were

THE POST-TRANSITION METALS AND METALLOIDS

IIIA	IVA	VA	VIA
B			
Al	**Si**		
Ga	**Ge**	**As**	
In	*Sn*	**Sb**	**Te**
Tl	*Pb*	*Bi*	**Po**

Note: Post-transition metals are in italics. Metalloids are in bold type.

discovered, and where they are found on Earth. It also details how these elements are used by humans and the resulting benefits and challenges to society, health, and the environment.

Chapters 1 through 5 cover the post-transition metals. The first chapter discusses aluminum, one of the most important materials for construction of vehicles, buildings, and medical and sports equipment. Unfortunately, aluminum smelters consume huge amounts of energy, and carbon dioxide emissions from this industry need to be addressed. Conservative estimates indicate that about five tons of CO_2 are emitted for every ton of pure aluminum separated.

Chapters 2 and 3 discuss gallium, indium, and thallium—elements that find important uses in the electronics and semiconductor industries. Thallium-doped lead-telluride may turn out to be a crucial breakthrough for transferring heat energy to electrical energy.

Chapters 4 and 5 examine tin, lead, and bismuth. Tin has been important for millennia in the production of bronze and pewter, both alloys of tin and copper. Tin and lead alloy to make solder. Tin has many household uses, whereas the use of lead in homes is highly discouraged due to its toxicity to humans. Lead and lead-bismuth coolants, however, are used in nuclear reactors. In addition, lead is a natural radiation shield.

Chapters 6 through 9 cover the metalloids. The subject of chapter 6 is boron, an essential nutrient for green algae and other plants. Boric acid and borax have been used for centuries as antiseptics. A new form of boron—"boron boride"—produced at very high pressure in the laboratory, was discovered in 2009.

Chapter 7 investigates two more elements that are essential in the electronics and semiconductor industries—silicon and germanium. The principal use of germanium is in transistors. Silicon, one of the most ubiquitous of elements on Earth, has been used in glassmaking for millennia, yet glassmaking remains a skill that is more art than science.

Chapters 8 and 9 explore some of the more poisonous elements—arsenic, antimony, and polonium—as well as the rather exotic but less hazardous element, tellurium. Several major uses of arsenic make use of its toxicity. White arsenic is used in the manufacture of insecticides, weed killers, and rodenticides. Antimony's major use is as a flame retardant for plastics and fabrics. Tellurium is an element of choice in solar cells because its conductivity increases when exposed to light.

Chapter 10 presents possible future developments that involve the post-transition metals and metalloids.

Metals and Metalloids will provide the reader, whether student or scientist, with an up-to-date understanding regarding each of the post-transition metals and metalloids—where they came from, how they fit into our current technological society, and where they may lead us.

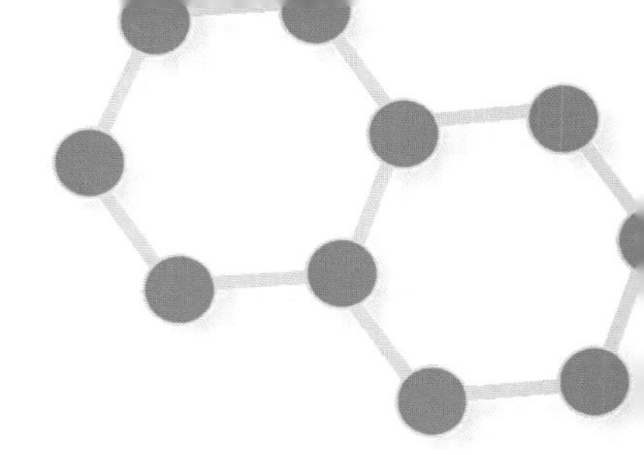

Overview: Chemistry and Physics Background

What *is* an element? To the ancient Greeks, everything on Earth was made from only four elements—earth, air, fire, and water. Celestial bodies—the Sun, moon, planets, and stars—were made of a fifth element: ether. Only gradually did the concept of an element become more specific.

An important observation about nature was that substances can change into other substances. For example, wood burns, producing heat, light, and smoke and leaving ash. Pure metals like gold, copper, silver, iron, and lead can be smelted from their ores. Grape juice can be fermented to make wine and barley fermented to make beer. Food can be cooked; food can also putrefy. The baking of clay converts it into bricks and pottery. These changes are all examples of chemical reactions. Alchemists' careful observations of many chemical reactions greatly helped them to clarify the differences between the most elementary substances ("elements") and combinations of elementary substances ("compounds" or "mixtures").

Elements came to be recognized as simple substances that cannot be decomposed into other even simpler substances by chemical reactions. Some of the elements that had been identified by the Middle Ages are easily recognized in the periodic table because they still have chemical symbols that come from their Latin names. These elements are listed in the table on page xviii.

The Russian chemist Dmitri Mendeleev created the periodic table of the elements in the late 1800s. *(SPL/Photo Researchers, Inc.)*

Modern atomic theory began with the work of the English chemist John Dalton in the first decade of the 19th century. As the concept of the atomic composition of matter developed, chemists began to define elements as simple substances that contain only one kind of atom. Because scientists in the 19th century lacked any experimental apparatus capable

ELEMENTS KNOWN TO ANCIENT PEOPLE

Iron: Fe ("ferrum")	Copper: Cu ("cuprum")
Silver: Ag ("argentum")	Gold: Au ("aurum")
Lead: Pb ("plumbum")	Tin: Sn ("stannum")
Antimony: Sb ("stibium")	Mercury: Hg ("hydrargyrum")
*Sodium: Na ("natrium")	*Potassium: K ("kalium")
Sulfur: S ("sulfur")	

Note: *Sodium and potassium were not isolated as pure elements until the early 1800s, but some of their salts were known to ancient people.

of probing the structure of atoms, the 19th-century model of the atom was rather simple. Atoms were thought of as small spheres of uniform density; atoms of different elements differed only in their masses. Despite the simplicity of this model of the atom, it was a great step forward in our understanding of the nature of matter. Elements could be defined as simple substances containing only one kind of atom. Compounds are simple substances that contain more than one kind of atom. Because atoms have definite masses, and only whole numbers of atoms can combine to make molecules, the different elements that make up compounds are found in definite proportions by mass. (For example, a molecule of water contains one oxygen atom and two hydrogen atoms, or a mass ratio of oxygen-to-hydrogen of about 8:1.) Since atoms are neither created nor destroyed during ordinary chemical reactions ("ordinary" meaning in contrast to "nuclear" reactions), what happens in chemical reactions is that atoms are rearranged into combinations that differ from the original reactants, but in doing so, the total mass is conserved. Mixtures are combinations of elements that are not in definite proportions. (In salt water, for example, the salt could be 3 percent by mass, or 5 percent by mass, or many other possibilities; regardless of the percentage of salt, it would still be called "salt water.") Chemical reactions are not required to separate the components of mixtures; the components of mixtures can be separated by physical processes such as distillation, evaporation, or precipitation. Examples of elements, compounds, and mixtures are listed in the following table.

EXAMPLES OF ELEMENTS, COMPOUNDS, AND MIXTURES

ELEMENTS	COMPOUNDS	MIXTURES
Hydrogen	Water	Salt water
Oxygen	Carbon dioxide	Air
Carbon	Propane	Natural gas
Sodium	Table salt	Salt and pepper
Iron	Hemoglobin	Blood
Silicon	Silicon dioxide	Sand

The definition of an element became more precise at the dawn of the 20th century with the discovery of the proton. We now know that an atom has a small center called the "nucleus." In the nucleus are one or more protons, positively charged particles, the number of which determine an atom's identity. The number of protons an atom has is referred to as its "atomic number." Hydrogen, the lightest element, has an atomic number of 1, which means each of its atoms contains a single proton. The next element, helium, has an atomic number of 2, which means each of its atoms contain two protons. Lithium has an atomic number of 3, so its atoms have three protons, and so forth, all the way through the periodic table. Atomic nuclei also contain neutrons, but atoms of the same element can have different numbers of neutrons; we call atoms of the same element with different number of neutrons "isotopes."

There are roughly 92 naturally occurring elements—hydrogen through uranium. Of those 92, two elements, technetium (element 43) and promethium (element 61), may once have occurred naturally on Earth, but the atoms that originally occurred on Earth have decayed away, and those two elements are now produced artificially in nuclear reactors. In fact, technetium is produced in significant quantities because of its daily use by hospitals in nuclear medicine. Some of the other first 92 elements—polonium, astatine, and francium, for example—are so radioactive that they exist in only tiny amounts. All of the elements with atomic numbers greater than 92—the so-called transuranium elements—are all produced artificially in nuclear reactors or particle accelerators. As of the writing of this book, the discoveries of the elements through number 118 have all been reported. The discoveries of elements with atomic numbers greater than 111 have not yet been confirmed, so those elements have not yet been named.

When the Russian chemist Dmitri Mendeleev (1834–1907) developed his version of the periodic table in 1869, he arranged the elements known at that time in order of *atomic mass* or *atomic weight* so that they fell into columns called *groups* or *families* consisting of elements with similar chemical and physical properties. By doing so, the rows exhibit periodic trends in properties going from left to right across the table, hence the reference to rows as *periods* and name "periodic table."

Mendeleev's table was not the first periodic table, nor was Mendeleev the first person to notice *triads* or other groupings of elements with similar properties. What made Mendeleev's table successful and the one we use today are two innovative features. In the 1860s, the concept of *atomic number* had not yet been developed, only the concept of atomic mass. Elements were always listed in order of their atomic masses, beginning with the lightest element, hydrogen, and ending with the heaviest element known at that time, uranium. Gallium and germanium, however, had not yet been discovered. Therefore, if one were listing the known elements in order of atomic mass, arsenic would follow zinc, but that would place arsenic between aluminum and indium. That does not make sense because arsenic's properties are much more like those of phosphorus and antimony, not like those of aluminum and indium.

To place arsenic in its "proper" position, Mendeleev's first innovation was to leave two blank spaces in the table after zinc. He called the first element eka-aluminum and the second element eka-silicon, which

Mendeleev's Periodic Table (1871)

Period \ Group	I	II	III	IV	V	VI	VII	VIII
1	H=1							
2	Li=7	Be=9.4	B=11	C=12	N=14	O=16	F=19	
3	Na=23	Mg=24	Al=27.3	Si=28	P=31	S=32	Cl=35.5	
4	K=39	Ca=40	?=44	Ti=48	V=51	Cr=52	Mn=55	Fe=56, Co=59 Ni=59
5	Cu=63	Zn=65	?=68	?=72	As=75	Se=78	Br=80	
6	Rb=85	Sr=87	?Yt=88	Zr=90	Nb=94	Mo=96	?=100	Ru=104, Rh=104 Pd=106
7	Ag=108	Cd=112	In=113	Sn=118	Sb=122	Te=125	J=127	
8	Cs=133	Ba=137	?Di=138	?Ce=140				
9								
10			?Er=178	?La=180	Ta=182	W=184		Os=195, Ir=197 Pt=198
11	Au=199	Hg=200	Tl=204	Pb=207	Bi=208			
12				Th=231		U=240		

Dmitri Mendeleev's 1871 periodic table. The elements listed are the ones that were known at that time, arranged in order of increasing relative atomic mass. Mendeleev predicted the existence of elements with masses of 44, 68, and 72. His predictions were later shown to have been correct.

he said corresponded to elements that had not yet been discovered but whose properties would resemble the properties of aluminum and silicon, respectively. Not only did Mendeleev predict the elements' existence, he also estimated what their physical and chemical properties should be in analogy to the elements near them. Shortly afterward, these two elements were discovered and their properties were found to be very close to what Mendeleev had predicted. Eka-aluminum was called *gallium* and eka-silicon was called *germanium*. These discoveries validated the predictive power of Mendeleev's arrangement of the elements and demonstrated that Mendeleev's periodic table could be a predictive tool, not just a compendium of information that people already knew.

The second innovation Mendeleev made involved the relative placement of tellurium and iodine. If the elements are listed in strict order of their atomic masses, then iodine should be placed before tellurium, since iodine is lighter. That would place iodine in a group with sulfur and selenium and tellurium in a group with chlorine and bromine, an arrangement that does not work for either iodine or tellurium. Therefore, Mendeleev rather boldly reversed the order of tellurium and iodine so that tellurium falls below selenium and iodine falls below bromine. More than 40 years later, after Mendeleev's death, the concept of atomic number was introduced, and it was recognized that elements should be listed in order of atomic number, not atomic mass. Mendeleev's ordering was thus vindicated, since tellurium's atomic number is one less than iodine's atomic number. Before he died, Mendeleev was considered for the Nobel Prize, but did not receive sufficient votes to receive the award despite the importance of his insights.

THE PERIODIC TABLE TODAY

All of the elements in the first 12 groups of the periodic table are referred to as *metals*. The first two groups of elements on the left-hand side of the table are the *alkali metals* and the *alkaline earth metals*. All of the alkali metals are extremely similar to each other in their chemical and physical properties, as, in turn, are all of the alkaline earths to each other. The 10 groups of elements in the middle of the periodic table are *transition metals*. The similarities in these groups are not as strong as those in the first two groups, but still satisfy the general trend of similar chemical and physical properties. The transition metals in the last row are not found in nature but have been synthesized artificially. The metals that follow the transition metals are called post-transition metals.

The so-called *rare earth elements*, which are all metals, usually are displayed in a separate block of their own located below the rest of the periodic table. The elements in the first row of rare earths are called *lanthanides* because their properties are extremely similar to the properties of lanthanum. The elements in the second row of rare earths are called *actinides* because their properties are extremely similar to the properties of actinium. The actinides following uranium are called *transuranium elements* and are not found in nature but have been produced artificially.

The far right-hand six groups of the periodic table—the remaining *main group elements*—differ from the first 12 groups in that more than one kind of element is found in them; in this part of the table we find metals, all of the *metalloids* (or *semimetals*), and all of the *nonmetals*. Not counting the artificially synthesized elements in these groups (elements having atomic numbers of 113 and above and that have not yet been named), these six groups contain seven metals, eight metalloids, and 16 nonmetals. Except for the last group—the *noble gases*—each individual group has more than just one kind of element. In fact, sometimes nonmetals, metalloids, and metals are all found in the same column, as are the cases with group IVB (C, Si, Ge, Sn, and Pb) and also with group VB (N, P, As, Sb, and Bi). Although similarities in chemical and physical properties are present within a column, the differences are often more striking than the similarities. In some cases, elements in the same column do have very similar chemistry. Triads of such elements include three of the *halogens* in group VIIB—chlorine, bromine, and iodine; and three group VIB elements—sulfur, selenium, and tellurium.

ELEMENTS ARE MADE OF ATOMS

An atom is the fundamental unit of matter. In ordinary chemical reactions, atoms cannot be created or destroyed. Atoms contain smaller *subatomic* particles: protons, neutrons, and electrons. Protons and neutrons are located in the *nucleus*, or center, of the atom and are referred to as *nucleons*. Electrons are located outside the nucleus. Protons and neutrons are comparable in mass and significantly more massive than electrons. Protons carry positive electrical charge. Electrons carry negative charge. Neutrons are electrically neutral.

The identity of an element is determined by the number of protons found in the nucleus of an atom of the element. The number of protons is called an element's atomic number, and is designated by the letter *Z*. For hydrogen, Z = 1, and for helium, Z = 2. The heaviest naturally occurring element is uranium, with Z = 92. The value of Z is 118 for the heaviest element that has been synthesized artificially.

Atoms of the same element can have varying numbers of neutrons. The number of neutrons is designated by the letter *N*. Atoms of the same element that have different numbers of neutrons are called *isotopes* of that element. The term *isotope* means that the atoms occupy the same place in the periodic table. The sum of an atom's protons and neutrons is called the atom's *mass number.* Mass numbers are dimensionless whole numbers designated by the letter *A* and should not be confused with an atom's *mass,* which is a decimal number expressed in units such as grams. Most elements on Earth have more than one isotope. The average mass number of an element's isotopes is called the element's atomic mass or atomic weight.

The standard notation for designating an atom's atomic and mass numbers is to show the atomic number as a subscript and the mass number as a superscript to the left of the letter representing the element. For example, the two naturally occurring isotopes of hydrogen are written $^{1}_{1}H$ and $^{2}_{1}H$.

For atoms to be electrically neutral, the number of electrons must equal the number of protons. It is possible, however, for an atom to gain or lose electrons, forming *ions.* Metals tend to lose one or more electrons to form positively charged ions (called *cations*); nonmetals are more likely to gain one or more electrons to form negatively charged ions (called *anions*). Ionic charges are designated with superscripts. For example, a calcium ion is written as Ca^{2+}; a chloride ion is written as Cl^{-}.

THE PATTERN OF ELECTRONS IN AN ATOM

During the 19th century, when Mendeleev was developing his periodic table, the only property that was known to distinguish an atom of one element from an atom of another element was relative mass. Knowledge of atomic mass, however, did not suggest any relationship between an element's mass and its properties. It took several discoveries—among them that of the electron in 1897 by the British physicist John Joseph

("J. J.") Thomson, *quanta* in 1900 by the German physicist Max Planck, the wave nature of matter in 1923 by the French physicist Louis de Broglie, and the mathematical formulation of the quantum mechanical model of the atom in 1926 by the German physicists Werner Heisenberg and Erwin Schrödinger (all of whom collectively illustrate the international nature of science)—to elucidate the relationship between the structures of atoms and the properties of elements.

The number of protons in the nucleus of an atom defines the identity of that element. Since the number of electrons in a neutral atom is equal to the number of protons, an element's atomic number also reveals how many electrons are in that element's atoms. The electrons occupy regions of space that chemists and physicists call *shells.* The shells are further divided into regions of space called *subshells.* Subshells are related to angular momentum, which designates the shape of the electron orbit space around the nucleus. Shells are numbered 1, 2, 3, 4, and so forth (in theory out to infinity). In addition, shells may be designated by letters: The first shell is the K-shell, the second shell the L-shell, the third the M-shell, and so forth. Subshells have letter designations, s, p, d, and f being the most common. The *n*th shell has *n* possible subshells. Therefore, the first shell has only an s subshell, designated 1s; the second shell has both s and p subshells (2s and 2p); the third shell 3s, 3p, and 3d; and the fourth shell 4s, 4p, 4d, and 4f. (This pattern continues for higher-numbered shells, but this is enough for now.)

An s subshell is spherically symmetric and can hold a maximum of 2 electrons. A p subshell is dumbbell-shaped and holds 6 electrons, a d subshell 10 electrons, and an f subshell 14 electrons, with increasingly complicated shapes.

As the number of electrons in an atom increases, so does the number of shells occupied by electrons. In addition, because electrons are all negatively charged and tend to repel each other *electrostatically,* as the number of the shell increases, the size of the shell increases, which means that electrons in higher-numbered shells are located, on the average, farther from the nucleus. Inner shells tend to be fully occupied with the maximum number of electrons they can hold. The electrons in the outermost shell, which is likely to be only partially occupied, will determine that atom's properties.

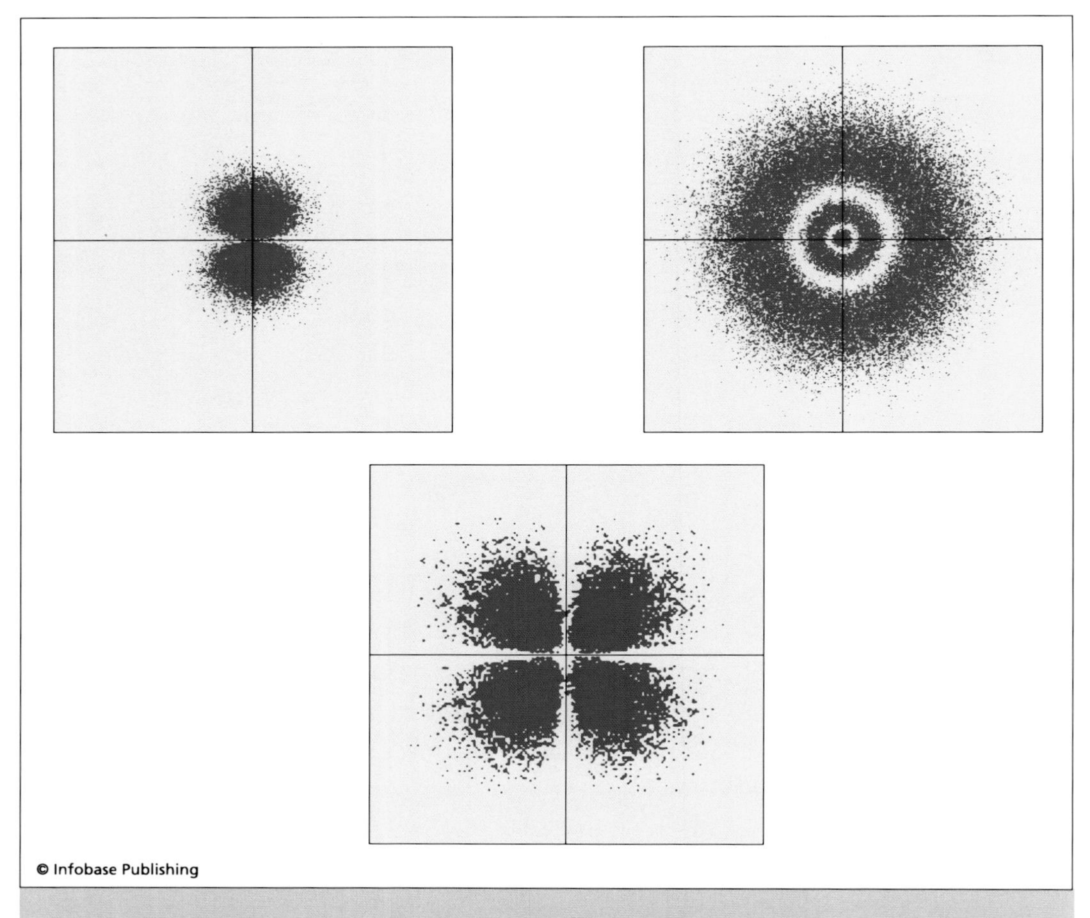

Some hydrogen wavefunction distributions for electrons in various excited states

Physicists and chemists use *electronic configurations* to designate which subshells in an atom are occupied by electrons as well as how many electrons are in each subshell. For example, nitrogen is element number 7, so it has seven electrons. Nitrogen's electronic configuration is $1s^2 2s^2 2p^3$; a superscript designates the number of electrons that occupy a subshell. The first shell is fully occupied with its maximum of two electrons. The second shell can hold a maximum of eight electrons, but it is only partially occupied with just five electrons—two in the 2s subshell and three in the 2p. Those five outer electrons determine nitrogen's properties. For a heavy element like tin (Sn), electronic configurations can be quite complex. Tin's configuration is $1s^2 2s^2 2p^6 3s^2 3p^6 4s^2 3d^{10} 4p^6 5s^2 4d^{10} 5p^2$ but is more commonly written in the shorthand notation [Kr] $5s^2 4d^{10} 5p^2$, where [Kr] represents the electron configuration pattern for the noble gas krypton. (The pattern continues in this way for shells with higher numbers.) The important thing to notice about tin's configuration is that all of the shells except the last one are fully occupied. The fifth shell can hold 32 electrons, but in tin there are only four electrons in the fifth shell. The outer electrons determine an element's properties. The table on page xxviii illustrates the electronic configurations for nitrogen and tin.

ATOMS ARE HELD TOGETHER WITH CHEMICAL BONDS

Fundamentally, a chemical bond involves either the sharing of two electrons or the transfer of one or more electrons to form ions. Two atoms of nonmetals tend to share pairs of electrons in what is called a *covalent bond*. By sharing electrons, the atoms remain more or less electrically neutral. However, when an atom of a metal approaches an atom of a nonmetal, the more likely event is the transfer of one or more electrons from the metal atom to the nonmetal atom. The metal atom becomes a positively charged ion and the nonmetal atom becomes a negatively charged ion. The attraction between opposite charges provides the force that holds the atoms together in what is called an *ionic bond*. Many chemical bonds are also intermediate in nature between covalent and ionic bonds and have characteristics of both types of bonds.

IN CHEMICAL REACTIONS, ATOMS REARRANGE TO FORM NEW COMPOUNDS

When a substance undergoes a *physical change*, the substance's name does not change. What may change is its temperature, its length, its *physical state* (whether it is a solid, liquid, or gas), or some other characteristic, but it is still the same substance. On the other hand, when a substance undergoes a *chemical change*, its name changes; it is a different substance. For example, water can decompose into hydrogen gas and oxygen gas, each of which has substantially different properties from water, even though water is composed of hydrogen and oxygen atoms.

In chemical reactions, the atoms themselves are not changed. Elements (like hydrogen and oxygen) may combine to form compounds

ELECTRONIC CONFIGURATIONS FOR NITROGEN AND TIN

ELECTRONIC CONFIGURATION OF NITROGEN (7 ELECTRONS)				ELECTRONIC CONFIGURATION OF TIN (50 ELECTRONS)			
Energy Level	Shell	Subshell	Number of Electrons	Energy Level	Shell	Subshell	Number of Electrons
1	K	s	2	1	K	s	2
2	L	s	2	2	L	s	2
		p	3			p	6
			7	3	M	s	2
						p	6
						d	10
				4	N	s	2
						p	6
						d	10
				5	O	s	2
						p	2
							50

(like water), or compounds can be decomposed into their elements. The atoms in compounds can be rearranged to form new compounds whose names and properties are different from the original compounds. Chemical reactions are indicated by writing chemical equations such as the equation showing the decomposition of water into hydrogen and oxygen: $2\ H_2O\ (l) \rightarrow 2\ H_2\ (g) + O_2\ (g)$. The arrow indicates the direction in which the reaction proceeds. The reaction begins with the *reactants* on the left and ends with the *products* on the right. We sometimes designate the physical state of a reactant or product in parentheses—*s* for solid, *l* for liquid, *g* for gas, and *aq* for *aqueous* solution (in other words, a solution in which water is the solvent).

IN NUCLEAR REACTIONS THE NUCLEI OF ATOMS CHANGE

In ordinary chemical reactions, chemical bonds in the reactant species are broken, the atoms rearrange, and new chemical bonds are formed in the product species. These changes only affect an atom's electrons; there is no change to the nucleus. Hence there is no change in an element's identity. On the other hand, nuclear reactions refer to changes in an atom's nucleus (whether or not there are electrons attached). In most nuclear reactions, the number of protons in the nucleus changes, which means that elements are changed, or transmuted, into different elements. There are several ways in which *transmutation* can occur. Some transmutations occur naturally, while others only occur artificially in nuclear reactors or particle accelerators.

The most familiar form of transmutation is *radioactive decay,* a natural process in which a nucleus emits a small particle or *photon* of light. Three common modes of decay are labeled *alpha, beta,* and *gamma* (the first three letters of the Greek alphabet). Alpha decay occurs among elements at the heavy end of the periodic table, basically elements heavier than lead. An alpha particle is a nucleus of helium 4 and is symbolized as ${}^{4}_{2}He$ or α. An example of alpha decay occurs when uranium 238 emits an alpha particle and is changed into thorium 234 as in the following reaction: ${}^{238}_{92}U \rightarrow {}^{4}_{2}He + {}^{234}_{90}Th$. Notice that the parent isotope, U-238, has 92 protons, while the daughter isotope, Th-234, has only 90 protons. The decrease in the number of protons means a change in the identity of the element. The mass number also decreases.

Any element in the periodic table can undergo beta decay. A beta particle is an electron, commonly symbolized as β^- or e^-. An example of beta decay is the conversion of cobalt 60 into nickel 60 by the following reaction: ${}^{60}_{27}Co \rightarrow {}^{60}_{28}Ni + e^-$. The atomic number of the daughter isotope is one greater than that of the parent isotope, which maintains charge balance. The mass number, however, does not change.

In gamma decay, photons of light (symbolized by γ) are emitted. Gamma radiation is a high-energy form of light. Light carries neither mass nor charge, so the isotope undergoing decay does not change identity; it only changes its energy state.

Elements also are transmuted into other elements by nuclear *fission* and *fusion*. Fission is the breakup of very large nuclei (at least as heavy as uranium) into smaller nuclei, as in the fission of U-236 in the following reaction: ${}^{236}_{92}U \rightarrow {}^{94}_{36}Kr + {}^{139}_{56}Ba + 3n$, where n is the symbol for a neutron (charge = 0, mass number = +1). In fusion, nuclei combine to form larger nuclei, as in the fusion of hydrogen isotopes to make helium. Energy may also be released during both fission and fusion. These events may occur naturally—fusion is the process that powers the Sun and all other stars—or they may be made to occur artificially.

Elements can be transmuted artificially by bombarding heavy target nuclei with lighter projectile nuclei in reactors or accelerators. The transuranium elements have been produced that way. Curium, for example, can be made by bombarding plutonium with alpha particles. Because the projectile and target nuclei both carry positive charges, projectiles must be accelerated to velocities close to the speed of light to overcome the force of repulsion between them. The production of successively heavier nuclei requires more and more energy. Usually, only a few atoms at a time are produced.

ELEMENTS OCCUR WITH DIFFERENT RELATIVE ABUNDANCES

Hydrogen overwhelmingly is the most abundant element in the universe. Stars are composed mostly of hydrogen, followed by helium and only very small amounts of any other element. Relative abundances of elements can be expressed in parts per million, either by mass or by numbers of atoms.

On Earth, elements may be found in the lithosphere (the rocky, solid part of Earth), the hydrosphere (the aqueous, or watery, part of Earth), or the atmosphere. Elements such as the noble gases, the rare earths, and commercially valuable metals like silver and gold occur in only trace quantities. Others, like oxygen, silicon, aluminum, iron, calcium, sodium, hydrogen, sulfur, and carbon are abundant.

HOW NATURALLY OCCURRING ELEMENTS HAVE BEEN DISCOVERED

For the elements that occur on Earth, methods of discovery have been varied. Some elements—like copper, silver, gold, tin, and lead—have been known and used since ancient or even prehistoric times. The origins of their early metallurgy are unknown. Some elements, like phosphorus, were discovered during the Middle Ages by alchemists who recognized that some mineral had an unknown composition. Sometimes, as in the case of oxygen, the discovery was by accident. In other instances—as in the discoveries of the alkali metals, alkaline earths, and lanthanides—chemists had a fairly good idea of what they were looking for and were able to isolate and identify the elements quite deliberately.

To establish that a new element has been discovered, a sample of the element must be isolated in pure form and subjected to various chemical and physical tests. If the tests indicate properties unknown in any other element, it is a reasonable conclusion that a new element has been discovered. Sometimes there are hazards associated with isolating a substance whose properties are unknown. The new element could be toxic, or so reactive that it can explode, or extremely radioactive. During the course of history, attempts to isolate new elements or compounds have resulted in more than just a few deaths.

HOW NEW ELEMENTS ARE MADE

Some elements do not occur naturally, but can be synthesized. They can be produced in nuclear reactors, from collisions in particle accelerators, or can be part of the *fallout* from nuclear explosions. One of the elements most commonly made in nuclear reactors is technetium. Relatively large quantities are made every day for applications in nuclear medicine. Sometimes, the initial product made in an accelerator is a heavy element whose atoms have very short *half-lives* and undergo radioactive decay. When the atoms decay, atoms of elements lighter than the parent atoms are produced. By identifying the daughter atoms, scientists can work backward and correctly identify the parent atoms from which they came.

The major difficulty with synthesizing heavy elements is the number of protons in their nuclei ($Z > 92$). The large amount of positive charge makes the nuclei unstable so that they tend to disintegrate either by radioactive decay or *spontaneous fission*. Therefore, with the exception of a few transuranium elements like plutonium (Pu) and americium (Am), most artificial elements are made only a few atoms at a time and so far have no practical or commercial uses.

THE METALS AND METALLOIDS SECTION OF THE PERIODIC TABLE

The metals discussed in this book are aluminum, gallium, indium, thallium, tin, lead, and bismuth. Technically, these are termed the *post-transition metals* because they follow the transition metals when reading the periodic table from left to right. For brevity, however, in the title of this book they are simply called "metals." To the right of this group in

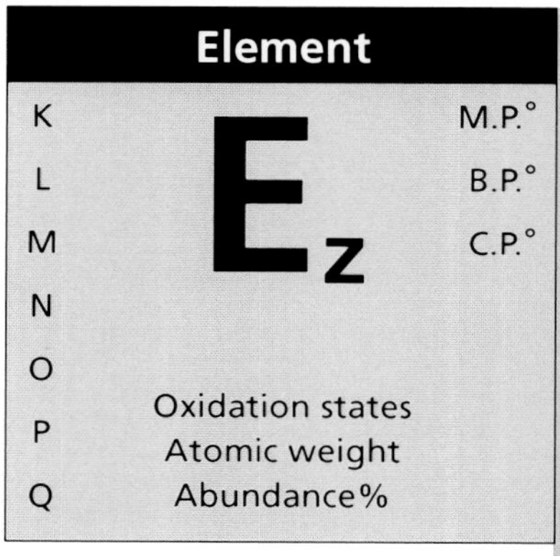

Element

K L M N O P Q

E_Z

M.P.° B.P.° C.P.°

Oxidation states
Atomic weight
Abundance%

Information box key. E represents the element's letter notation (for example, H = hydrogen), with the Z subscript indicating proton number. Orbital shell notations appear in the column on the left. For elements that are not naturally abundant, the mass number of the longest-lived isotope is given in brackets. The abundances (atomic %) are based on meteorite and solar wind data. The melting point (M.P.), boiling point (B.P.), and critical point (C.P.) temperatures are expressed in Celsius. Sublimation and critical temperatures are indicated by s and t.

the periodic table are the metalloids: boron, silicon, germanium, arsenic, antimony, tellurium, and polonium.

The book includes two groups of elements that are some of society's most industrially important substances. Tin and lead are familiar metals that have played important roles in civilization for thousands of years. Aluminum and silicon are particularly important in today's economy—aluminum as a lightweight structural material and silicon as the building block of modern electronics. In this book, readers will learn about the important properties of post-transition metals and the metalloids and how these elements are useful in everyday life.

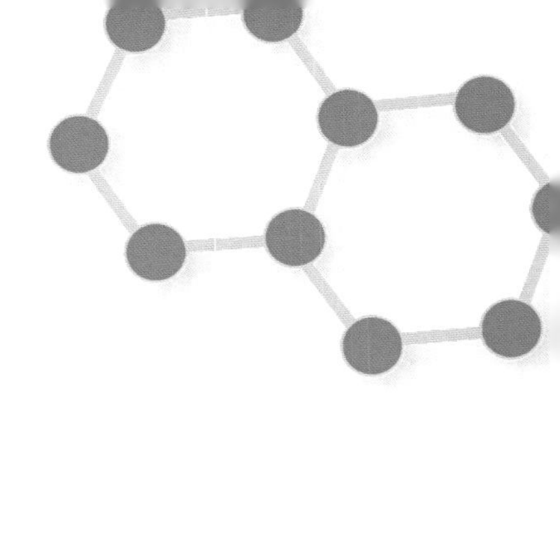

PART I

POST-TRANSITION METALS

INTRODUCTION TO THE POST-TRANSITION METALS

The post-transition metals are the elements located in the columns to the right of the transition elements, that is, to the right of the column that contains zinc, cadmium, and mercury. The post-transition metals are aluminum (Al), gallium (Ga), indium (In), thallium (Th), tin (Sn), lead (Pb), and bismuth (Bi). Aluminum is usually included with the post-transition metals, despite the fact that it does not exactly have the same characteristics.

To understand the chemistry of the elements, it is important to understand the concepts of *oxidation* and *reduction. Oxidation states* indicate the nature of the chemical bonding an atom exhibits in compounds or ions. (A pure element is defined as being in a zero oxidation state, indicating that it is not bonded to atoms of other elements.)

Positive oxidation states indicate either that an atom has lost electrons to form a positive ion or that it is bonded to atoms of other elements that attract electrons more strongly than it does (i.e., elements that are said to be more *electronegative*). Negative oxidation states indicate either that an atom has gained electrons to form a negative ion or that it is more electronegative than the atoms of other elements to which it is bonded.

An element's relative tendency to pull electrons away from atoms to which it is bonded is referred to as the element's electronegativity. Electronegativities are measured on a relative scale that has no units. Fluorine—the most electronegative element—is usually assigned an electronegativity of 4.0 while the remaining elements are assigned electronegativities greater than 0 but less than 4. Generally speaking, electronegativities are highest in the upper right-hand corner of the periodic table and lowest in the lower left-hand corner. The closer an element is to fluorine, the more electronegative that element is likely to be. An exception is the noble gases, which tend to be excluded from discussions of electronegativities because of their general tendency not to bond to other elements.

For example, consider the compound aluminum chloride ($AlCl_3$), which is characterized by covalent bonding. To say that chlorine is more electronegative than aluminum means that more negative charge is associated with the chlorine atoms, which places them in a negative oxidation state. That leaves the aluminum atom with relatively less negative charge, placing it in a positive oxidation state. As another example, carbon is in a positive oxidation state in carbon dioxide (CO_2) because oxygen atoms are more electronegative than carbon atoms are. On the other hand, carbon is in a negative oxidation state in methane (CH_4) because carbon atoms are more electronegative than hydrogen atoms are. In general, because so many chemical compounds contain oxygen or hydrogen, it is useful to remember that in compounds containing oxygen, elements other than the oxygen tend to be in positive oxidation states. In compounds containing hydrogen, with relatively few exceptions, elements other than the hydrogen tend to be in negative oxidation states.

In the formulas of *binary* compounds (compounds containing two elements), the less electronegative element usually is written first, followed by the more electronegative element. (Methane [CH_4] and ammonia [NH_3] are exceptions; carbon and nitrogen are both more electronegative than hydrogen. However, writing hydrogen first would suggest that CH_4 and NH_3 are acids, which they are not. Writing hydrogen last emphasizes that these compounds are not acids.) Since nonmetals are always more electronegative than metals are, the formulas of *salts* (which consist of a metal and a nonmetal) and *metal oxides* are always written with the metal first, then the nonmetal. Examples include sodium chloride (NaCl), calcium fluoride (CaF_2), and ferric oxide (Fe_2O_3).

In compounds or ions, metals exist only in positive oxidation states. Nonmetals and metalloids (also called *semimetals*), on the other hand, may exhibit both positive or negative oxidation states, which is an important distinction between metals, and nonmetals or metalloids. Elements or compounds also may be classified on the basis of their roles in *oxidation-reduction* reactions. If the oxidation state of an element increases during the course of a chemical reaction, the element is said to have been oxidized. If the oxidation state decreases, the element is said to have been reduced. Oxidation-reduction reactions are chemical reactions in which one element is oxidized and another element is reduced. In oxidation-reduction reactions, an *oxidizing agent* is a chemical substance that pulls electrons away from other elements; the electronegative aspect of many nonmetals explains why they are often good oxidizing agents. As the most electronegative element in the periodic table, fluorine (as gaseous F_2) is a good example of a powerful oxidizing agent. Fluorine oxidizes other elements by taking electrons from them, which reduces the neutral fluorine atoms and converts them into fluoride ions (F^-).

In contrast, a *reducing agent* is a chemical substance that gives up electrons to atoms of other elements. Neutral metals tend to be good reducing agents. An oxidizing agent itself undergoes reduction, and a reducing agent itself undergoes oxidation. Therefore, an oxidation-reduction reaction may also be thought of as the reaction between an oxidizing agent and a reducing agent.

It is useful to compare and contrast the properties of metals, metalloids, and nonmetals. The chemical and physical properties of most of the nonmetals are determined by the fact that they all have in common the same kind of outermost (or valence) electrons—"p" electrons. Hydrogen and helium are the exceptions; their valence electrons are "s" electrons. An important property of nonmetals is that they commonly can exist in both positive and negative *oxidation states.* Likewise, metalloids tend to have chemical properties that are similar to the properties of nonmetals. Thus, metalloids, which—like most nonmetals—are found only in the "p" block, also commonly exist in both positive and negative oxidation states.

In contrast, due at least in part to the fact that there are many more metals than nonmetals on Earth, metals are located in all the blocks of the periodic table. An extremely important difference between metals and nonmetals is that—with only rare, rather artificial exceptions that will not be considered in this book—all of the metals exist only alone as neutral elements or in compounds or ions in positive oxidation states. An important implication of not existing in negative oxidation states is that it is safe to assume that metals do not form negatively charged ions. This property of metals means that they give up electrons rather easily, which is closely related to the fact that metals are conductors of electricity, in contrast to nonmetals, which are electrical insulators.

Looking for a moment at the distribution of metallic elements in the periodic table, it may be observed that the alkali and alkaline earth metals are located in the "s" block, meaning that their valence electrons are "s" electrons. The post-transition metals are located in the "p" block; therefore, their valence electrons are "p" electrons. The transition metals are located in the "d" block with "d" valence electrons. And finally, the lanthanides and actinides are located in the "f" block with "f" valence electrons. Different kinds and numbers of valence electrons give the elements in each block properties that differ from the properties of elements in the other blocks.

The chemical properties of the post-transition metals are aligned closely with the group numbers of the columns in which they are found. Aluminum, gallium, indium, and thallium are in group IIIA, which means, in general, that the chemistry of their "+3" ions tends to be dominant. The exception is thallium, for which the chemistry of the "+1" ion is more important. Tin and lead are in group IVA. Although their chemistry tends to be dominated by the "+2" ion in each case, the chemistry of tin and lead in their "+4" oxidation states is also very important. Bismuth is in group VA. The chemistry of its "+3" ion is more important, but bismuth does exist in compounds in the "+5" oxidation state.

Tin and lead have been important metals since the earliest times of human history. Bismuth was known during the Middle Ages. Otherwise, the discoveries and applications of the remaining post-transition metals tend to have occurred relatively recently in history. The group IIIA metals were discovered during the 19th century. Because they tend to be very difficult to extract from their ores, applications for them were only found during the 20th century. As an example of a group IIIA metal, aluminum definitely is the one that has found the most applications in modern industry. Therefore, the discussion of the post-transition metals will begin with aluminum—a very lightweight, corrosion resistant, and versatile element.

1

Aluminum

Aluminum (called "aluminium" in Britain and some of its former colonies)—element 13 in the periodic table—is a silvery metal with a density of 2.70 g/cm^3. It has only one isotope on Earth—$^{27}_{13}Al$. Aluminum is found as a *silicate* material (that is, in combination with silicon and oxygen) in almost all rocks and is the third most abundant element—and the most abundant metal—on Earth. Bauxite (mostly Al_2O_3) is the most important ore of aluminum. Another important ore is corundum, which also is a form of Al_2O_3 and is useful for its abrasive properties. Aluminum is an excellent conductor of electricity and heat. In addition, its surface is highly reflective. Of the light (85–90 percent) incident on an aluminum surface, most is reflected, rather than being absorbed. Because pure aluminum by itself is too soft to use as a structural material (in aircraft, for example, where its light weight is

THE BASICS OF ALUMINUM

Symbol: Al
Atomic number: 13
Atomic mass: 26.9815386
Electronic configuration: $[Ne]3s^2 3p^1$

T_{melt} = 1,221°F (660°C)
T_{boil} = 4,566°F (2,519°C)

Abundance
In Earth's crust 82,000 ppm
In seawater 0.005 ppm

Aluminum		
2	Al	660.32°
8	13	2519°
3		
	+3	
	26.9815386	
	0.000277%	

Isotope	Z	N	Relative Abundance
$^{27}_{13}Al$	13	14	100%

of utmost importance), other elements are alloyed with aluminum to make it strong, machinable, and formable into many useful shapes.

The corrosion resistance of aluminum is demonstrated by contrasting aluminum with iron. The "+3" ion is important in the chemistry of both aluminum and iron. The "+3" ion of iron (the "ferric" ion) readily forms ferric oxide (Fe_2O_3), which is the familiar reddish-brown substance commonly referred to as rust.

The following chemical reactions show rust's formation:

$$4\ Fe\ (s) + 3\ O_2\ (g) + 6\ H_2O\ (l) \rightarrow 4\ Fe(OH)_3\ (s);$$

$$2\ Fe(OH)_3\ (s) \rightarrow Fe_2O_3\ (s) + 3\ H_2O\ (l).$$

Ferric oxide is brittle and can be easily chipped off the surface, exposing fresh metal underneath to additional corrosion to the extent that holes (called *pits*) are formed.

The aluminum "+3" ion also forms an oxide, as shown by reactions analogous to the reactions for the formation of ferric oxide:

$$4\ Al\ (s) + 3\ O_2\ (g) + 6\ H_2O\ (l) \rightarrow 4\ Al(OH)_3\ (s);$$

$$2\ Al(OH)_3\ (s) \rightarrow Al_2O_3\ (s) + 3\ H_2O\ (l).$$

In contrast to ferric oxide, however, aluminum oxide (Al_2O_3) is very durable and tends to coat the surface, thus preventing additional corrosion. Aluminum's resistance to corrosion is responsible for the long lifetimes of aluminum cans in the environment. Unlike iron or steel cans, which in a moist environment might disintegrate in only a few years, aluminum cans can last a century in similar environments.

Aluminum metal also is much more expensive to obtain from its ores than iron metal is. This fact is what makes recycling aluminum products so extremely important. Not only does recycling mean that less aluminum ends up in landfills, it simply makes good economic sense to recycle aluminum. It takes only 5 percent as much energy to recycle an aluminum product and fashion it into a new product as it takes to extract aluminum ore from the ground, reduce the ore to the metal, and then fashion the metal into a useful product. Using aluminum to make soft-drink cans may be extremely convenient, but the convenience is justified only if the cans are recycled.

Aluminum alloys are easily formed into useful shapes. *(Adisa/ Shutterstock)*

THE ASTROPHYSICS OF ALUMINUM

The aluminum that exists in the universe today is largely a product of chaotic collisions that occur in supernova explosions, which then distribute the aluminum *isotopes* into interstellar space. One particular isotope—aluminum 26—is proving useful in establishing a timescale for interstellar events. Aluminum 26 is radioactive. From the time it is created, it begins to disappear via the following decay process.

$$^{26}_{13}\mathrm{Al} \rightarrow {}^{26}_{12}\mathrm{Mg} + \gamma + e^{+}$$

where γ indicates a gamma ray and e^{+} a positron.

By contrast, aluminum 27, which is also synthesized in supernovae, is extremely stable. As the ^{26}Al continues to decay with a *half-life* of 7.2×10^5 years, the isotopic ratio ^{26}Al/^{27}Al decreases accordingly, and, like carbon-dating, can be used to monitor the date of origin of the sample in question.

Stellar and solar system events of interest, however, generally occurred much farther in the past than the few million years it would have taken for all of the ^{26}Al to decay. Fortunately, the ^{26}Mg by-product is quite stable. While magnesium 26 is also produced in supernovae, any excess over its normal astrophysical abundance can only be attributed to the decay of aluminum 26. Because the ^{26}Mg abundance increases in direct proportion to the decay of ^{26}Al, current measurements of ^{26}Mg/^{27}Al in a medium correspond to the original ^{26}Al/^{27}Al ratio that existed at the time of formation of that sample—when the maximum amount of aluminum 26 existed.

The collapse of the gaseous cloud that resulted in the solar system about four and a half billion years ago is one astrophysical event about which the original ^{26}Al content informs science. The solar system contains an excess of ^{26}Mg over interstellar abundances. This indicates that a supernova must have occurred in the solar vicinity less than a few million years before the solar *nebula* formed.

Another area where ^{26}Al is important is in the understanding of how *asteroids* melt. Asteroids travel at fairly high speeds, but collisions cannot account for the observed evidence of past melting and resolidification. The decay of aluminum 26 provides an answer. When the *fission* occurs, the magnesium nucleus departs in the opposite direction

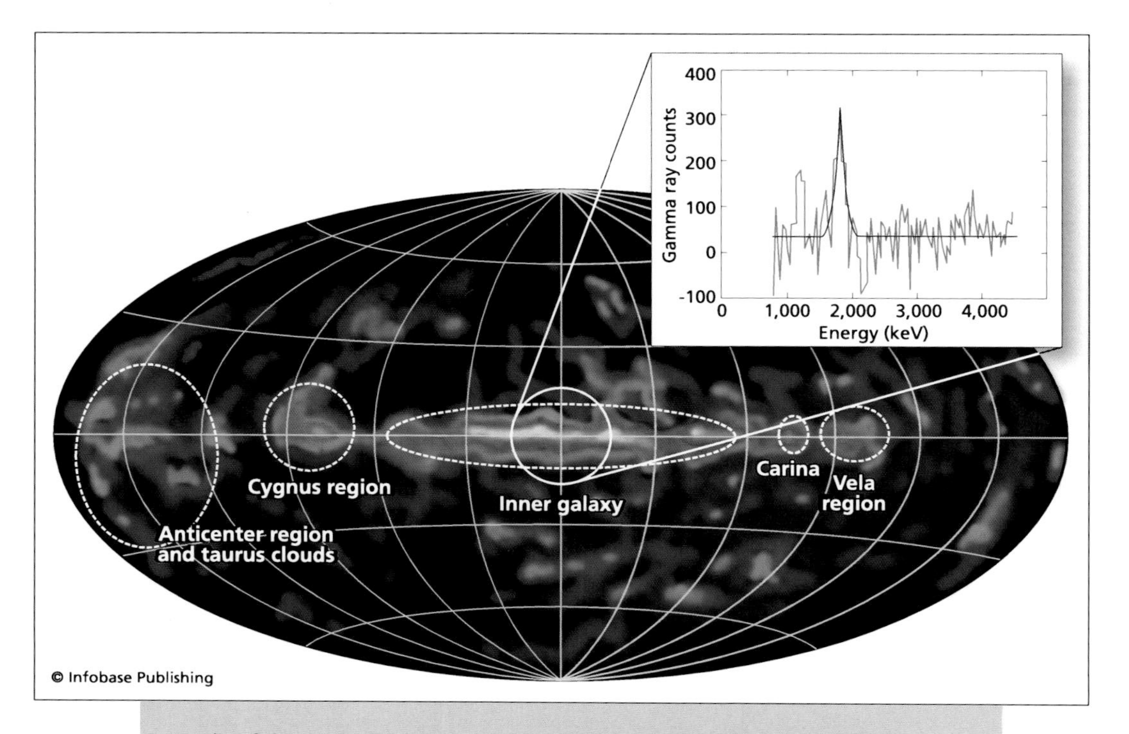

In the false color image shown here, regions of no aluminum emission are dark, while regions of green-red-yellow show increasingly bright regions of Al-26 emission. This map clearly shows that Al-26 emission is associated with regions of massive star formation in the galaxy, like the Carina Nebula, the Cygnus region, and the region near the center of the galaxy.

from the gamma ray. When embedded in a solid body like an asteroid, the motion of the magnesium nucleus creates heat energy that can not escape, but can melt the surrounding medium.

The importance of ^{26}Al as a *chronometer* is currently undisputed, but its origin is somewhat questionable. Recent studies show that supernovae may not be the only source of aluminum in the interstellar medium (ISM). Astronomers have observed ^{26}Al in the atmospheric spectra of high-mass stars, called *asymptotic giant branch* (AGB) stars. Aluminum 26 is most likely produced in the hydrogen-burning layer of such stars by the following reaction:

$$^{25}_{12}\mathrm{Mg} + {}^{1}_{1}\mathrm{p} \rightarrow {}^{26}_{13}\mathrm{Al}.$$

This would require a mixing process by which material from the outer envelope of the star is carried to the H-burning layer. This circulation is poorly understood, but would be effective at providing the ambient interstellar medium with aluminum 26 on the outflowing stellar wind.

Gamma-ray imaging is an ideal way to find the highest concentrations of ^{26}Al formation. The signature energy of 1.8 MeV from gamma rays emitted in the decay of ^{26}Al has been detected by satellite observatories. The highest emission density is associated with areas of massive star formation, and is particularly strong near the galactic center.

DISCOVERY AND NAMING OF ALUMINUM

Aluminum is the most abundant metal in Earth's crust. Today, aluminum is one of our most industrially important metals, and yet it was the 54th element to be discovered. Alum—from the mineral alunite, $KAl_3(SO_4)_2(OH)_6$—had been used for centuries in medicine and for dyeing cloth, but its composition had not been investigated. Aluminum hydroxide, or $Al(OH)_3$, was also used in papermaking, goldsmithing, bookbinding, preserving anatomical specimens, and in embalming.

The discoverer of aluminum, Hans Christian Ørsted (1777–1851) was born in Denmark. The son of a pharmacist, young Hans Christian learned some rudimentary chemistry as his father's assistant. Later, Hans Christian was able to attend the University of Copenhagen, where he studied science and medicine and became a doctor of medicine. After graduation, Ørsted became interested in physics and electricity, and he became a professor of physics at the University of Copenhagen in 1806.

Combining his interests in electricity and chemistry, Ørsted studied *electrochemistry*—the inducement of a chemical reaction by means of an electrical current. One of his first subjects of study was the composition of alumina. In 1825, by wafting chlorine gas over a mixture of hot charcoal and alumina, Ørsted prepared aluminum chloride, $AlCl_3$. He then reacted the aluminum chloride with an *amalgam* of potassium and mercury, resulting in an amalgam of aluminum and mercury. After distilling the mercury, Ørsted was left with a small sample of metallic aluminum. In recognition of this accomplishment, he is credited with aluminum's discovery. The name *aluminum* comes from the alumina from which it was prepared.

Ørsted did not continue his investigation of aluminum. That task fell to the German chemist Friedrich Wöhler (1800–82), who modified Ørsted's method of preparing aluminum and, in 1827, obtained a large enough sample that he could study aluminum's properties. Metallic aluminum still could not be produced in sufficient quantities, however, to be commercially useful. It was found that reacting aluminum chloride with sodium metal rather than with potassium yielded more product. Sodium, however, was more expensive than aluminum, so less expensive methods for producing sodium would have to be found before the manufacture of aluminum could become feasible by that method.

It was not until 1886 that a truly economical method of obtaining aluminum became available. In that year, Charles Martin Hall (1863–1914), a student at Oberlin College in Ohio, built a laboratory in his parents' woodshed. Inspired by his chemistry professor to find an economical method for manufacturing aluminum, Hall developed the process that bears his name of electrolytically decomposing aluminum oxide, resulting in appreciable quantities of aluminum globules. Hall

Inspired by his chemistry professor to find an economical method for manufacturing aluminum, Charles M. Hall developed the process, which bears his name, of electrolytically decomposing aluminum oxide. *(Alcoa)*

Paul-Louis-Toussaint Héroult of France independently developed the electrolytic process for decomposing aluminum oxide. *(Alcoa)*

went on to found the company that later became the Aluminum Company of America (ALCOA), today one of largest corporations in the United States. A statue commemorating Charles Hall stands today on the campus of Oberlin College.

Independently of Hall, Paul-Louis-Toussaint Héroult (1863–1914) of France simultaneously developed the same electrolytic process that Hall did. Today, the aluminum industry's method of obtaining aluminum from its ore is known as the Hall–Héroult process.

Hall's development of the electrochemical process for producing aluminum in his family's backyard woodshed has often been compared to the more modern work of Steven Jobs and Steve Wozniak, who built the first Apple® computer in the home of Jobs's parents. Just as Hall's entrepreneurial work resulted in ALCOA, a giant in the aluminum industry, Jobs's and Wozniak's work resulted in Apple, Inc., a giant in the personal computer industry.

THE CHEMISTRY OF ALUMINUM

With an electronic configuration of $1s^22s^22p^63s^23p^1$, aluminum—like all the elements in group IIIA—has three valence electrons. Therefore, in compounds, its only common oxidation state is "+3."

Aluminum is very easily oxidized. It dissolves rapidly in strong *acids* like hydrochloric acid (HCl) and sulfuric acid (H_2SO_4). Aluminum also dissolves in weak acids like acetic acid ($HC_2H_3O_2$) and in bases like sodium hydroxide (NaOH). Examples of these reactions are demonstrated by the following chemical equations:

$$2\ Al\ (s) + 6\ HCl\ (aq) \rightarrow 2\ AlCl_3\ (aq) + 3\ H_2\ (g);$$

$$2\ Al\ (s) + 2\ NaOH\ (aq) + 6\ H_2O\ (l) \rightarrow$$
$$2\ Na^+\ (aq) + 2\ Al(OH)_4^-\ (aq) + 3\ H_2\ (g).$$

Because aluminum reacts so readily with acids and bases found in nature, it is never found as the pure metal (except, of course, in trash). Aluminum is only found naturally combined with nonmetallic elements in the form of ores.

Chemical compounds that dissolve in both acids and bases are said to be *amphoteric*. In reactions with acids, aluminum forms soluble salts, which are all white in color. Examples are aluminum chloride ($AlCl_3$, as in the first equation above), aluminum nitrate ($Al(NO_3)_3$), and aluminum acetate ($Al(C_2H_3O_2)_3$). In reactions with weak bases, or in small amounts of strong bases, the aluminum ion precipitates as aluminum hydroxide ($Al(OH)_3$). In excess strong base, aluminum forms a *complex ion* with hydroxide—$Al(OH)_4^-$, as shown in the second equation above. When dry, aluminum hydroxide becomes aluminum oxide (Al_2O_3).

Anions, which are strong bases (such as the sulfide ion, S^{2-}), tend to *hydrolyze* in aqueous solutions to yield appreciable numbers of hydroxide ions (OH^-). Hydrolysis is illustrated by the example of sulfide ion in the following equation:

$$S^{2-}\ (aq) + H_2O\ (l) \rightarrow HS^-\ (aq) + OH^-\ (aq).$$

The result is that in a solution containing Al^{3+} and sodium sulfide (Na_2S), for example, the concentration of hydroxide ion (OH^-) is sufficient to precipitate $Al(OH)_3$ rather than Al_2S_3.

In analogy to aluminum metal itself, $Al(OH)_3$ also dissolves in both acidic and basic aqueous solutions, indicating its similar amphoteric nature. Examples of these reactions are shown in the following chemical equations:

$$Al(OH)_3\ (s) + 3\ HCl\ (aq) \rightarrow AlCl_3\ (aq) + 3\ H_2O\ (l);$$

$$Al(OH)_3\ (s) + NaOH\ (aq) \rightarrow Na^+\ (aq) + Al(OH)_4^-\ (aq).$$

Because aluminum ions form so few insoluble precipitates, the qualitative detection of aluminum in solutions requires a distinctive test. Most of the transition and post-transition metals readily precipitate as hydroxides in alkaline solutions. Among relatively common metals, however, aluminum, chromium, and zinc are amphoteric, meaning that their hydroxides redissolve at high *pHs* (i.e., high concentrations of hydroxide ion). Adding concentrated NaOH to a mixture of metal hydroxides will leave most solids intact, but will dissolve aluminum, chromium, and zinc hydroxides with the formation of the $Al(OH)_4^-$, $Cr(OH)_4^-$, and $Zn(OH)_4^{2-}$ ions, respectively. Solutions containing these ions can be acidified just slightly to reprecipitate $Al(OH)_3$, $Cr(OH)_3$, and $Zn(OH)_2$. At this point, the addition of a *reagent* called *aluminon* will result in the $Al(OH)_3$ turning cherry red (due to a mixture of the precipitate and a coating of aluminon), which is the definitive test for the presence of aluminum.

In the *thermite* process, aluminum metal will vigorously reduce various metal oxides to their neutral metals. An example is the reduction of iron in the reaction between aluminum and ferric oxide, as shown by the following chemical equation:

$$2\ Al\ (s) + Fe_2O_3\ (s) \rightarrow Al_2O_3\ (s) + 2\ Fe\ (s).$$

This reaction is very *exothermic,* releasing 5.3 kilojoules (kJ) of heat per gram of Fe_2O_3 reduced.

In sufficient dosages, the aluminum ion is a *neurotoxin* to humans and other animals; in other words, aluminum can adversely affect the nervous system. As a general rule, in nature aluminum ions tend to be tightly bound in silicate rocks, where animals do not come into contact with them. Environmental acids and bases, however, can dissolve the rocks and release aluminum ions into soils, streams, and lakes. The potentially harmful effects of aluminum ions on aquatic organisms is one of the serious consequences of *acid rain,* which is defined as rainfall with an acidity greater than unpolluted rainfall.

Compounds that contain aluminum are used in antiperspirants. Aluminum reacts with sweat to produce gels that block further sweating. Because of the concern of possible aluminum toxicity, however, some manufacturers offer alternative products that do not contain aluminum.

ENERGY CONSUMED IN ALUMINUM PRODUCTION

Aluminum is used in myriad products, from food and drink packaging to materials for airplanes. In order to supply this large amount of aluminum, *smelters* in the United States alone annually output about 5 billion kilograms in the form of *ingots.* This requires by far the largest energy consumption of any industry—about 15 kWh/kg of aluminum on average. (An average American household uses about 10,000 kWh per year, so the aluminum industry consumes as much energy annually as 7½ million households—approximately the number of homes in the state of New York, according to 2010 census data.)

Worldwide, aluminum production comprises 3 percent of all energy use, but in those countries with large bauxite concentrations or inexpensive electric power, the ratio is much higher. In Australia, 10 percent of all electricity goes into aluminum processing. In Iceland, where geothermal power is readily available (prompting Alcoa to establish a plant there), aluminum smelting consumes more energy than all other demands.

Industrial electrolysis—the procedure by which molten Al_2 is separated from the aluminum oxide (Al_2O_3) in bauxite ore—is carried out in what is called a *Hall-Héroult cell,* which operates at about 950°C (1,740°F). An important component of the energy waste in this process is the lining of the cell, which should provide protection against heat loss. Ceramics research may provide improvements in the near future. According to William Choate and John Greene, who prepared a special report for the U.S. Department of Energy in 2003, "The development of better, longer-lasting, ceramic materials for furnace linings is ongoing and will reduce the time required for furnace maintenance."

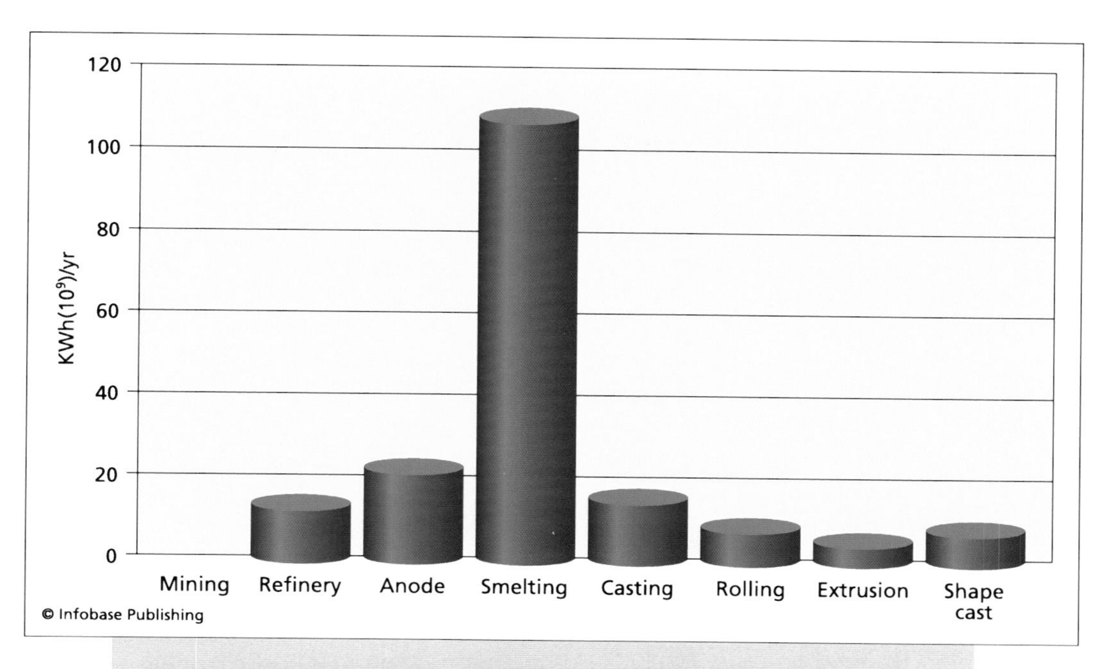

Aluminum smelters consume far more energy than other aluminum operations.

No improvements in furnace lining or bauxite ore refinement, however, will be able to provide the drastic reduction in energy use that accompanies simple recycling of aluminum products. This so-called secondary aluminum processing consumes a mere 5 percent of the energy needed to smelt ore. The promotion of aluminum recycling needs to be recognized as a crucial component in reducing global energy consumption, which could dramatically alleviate CO_2 emissions. Current aluminum recycling replaces only about 20 percent of annual aluminum consumption in the United States. Education is the key to this simple solution.

ALUMINUM SMELTER EMISSIONS

The huge power demand in aluminum processing is not the only environmental problem associated with this industry. The separation of pure aluminum from its aluminum oxide or from bauxite ore currently employs a method that pollutes the surrounding air, soil, and water with a variety of contaminants. Aluminum is separated from alumina in the following reaction:

$$2Al_2O_3 + 3C \rightarrow 4Al + 3CO_2$$

by *electrolysis*, where the *anode* is carbon and the cathode is aluminum. The medium in which electrolysis occurs is called the electrolytic solution or "*electrolyte*." In this case, the solution used is cryolite (Na_3AlF_6), with a proportion of aluminum fluoride (AlF_3) commonly added to optimize conditions for the procedure.

Circulation of the electrolyte allows gas bubbles to form, which then easily escape, carrying fluoride compounds and carbon dioxide (CO_2) into the atmosphere. The released carbon dioxide mixes with the atmosphere as a *greenhouse gas*. Conservative estimates indicate that about five tons of CO_2 are emitted for every ton of pure aluminum separated. The fluoride compounds generally settle to Earth to be absorbed by plants or soil. Consumption by livestock of forage crops that have absorbed excess fluoride has caused loss of animals to lameness and death from bone damage. Pollinating insects also suffer, and crop yields generally fall. White pine loss in some forests due to *fluorosis* has been observed in the proximity of aluminum smelters in Canada. Of great concern is the tendency for *biological magnification* of fluorides—the compounds become more concentrated as they move up the food chain. Additionally, some plants synthesize organic fluoride compounds like fluoroacetate and fluoricitrate from inorganic fluorides. These substances, which are extremely poisonous, have been detected in grasses as well as salad crops.

The electrolysis of alumina also produces perfluorocarbons as a result of reactions between the aluminum cathode and the electrolyte solution. The global-warming potential of these gases is thousands of times higher than that of carbon dioxide. One ton of hexafluoroethane (CF_6), for example, has the global warming effect of 9,000 tons of CO_2.

Unfortunately, pollution problems have arisen even in recycled aluminum processing because of paint or other coatings on aluminum scrap metal. These are difficult to separate from the metal and contain dioxins, furans, and other hazardous compounds.

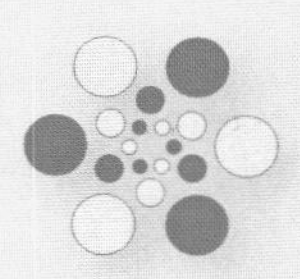

TRANSPARENT ALUMINA

Clear glass made from silica works well for windows in houses, but some material scientists have learned how to use aluminum oxide to make glass harder and to take advantage of optical characteristics like fluorescence and wavelength filtering.

Glass is a curious state of matter. It is not liquid or solid, but has properties of both. To make a glass, the trick is to prevent the substance from crystallizing into a regular solid, since the layers in crystal structures refract and reflect visible wavelengths rather than transmitting them. The relevant procedure requires cooling the molten liquid material fairly quickly—before it has the opportunity to form molecular crystal bonds.

While, in principle, any substance could be made into a glass, it is often not possible or economically feasible in practice, mostly because the necessary temperatures and pressures depend on the material and can be extreme in many cases. Scientists have recently found a way, however, to incorporate aluminum oxide with rare earth or calcium oxides to make a particularly resilient glass—one that does not easily deform upon impact. To deform a material uses energy, which means that momentum is not conserved in a collision. A material that minimizes such losses is useful in sporting equipment, such as golf clubs or baseball bats, designed to provide maximum rebound in collisions.

When allowed to crystallize, the alumina–rare earth oxide compounds are proving useful as extremely hard nanoscale ceramics. Transparent alumina films and fibers have also been produced by combining aluminum chloride with methyl cellulose, an organic composite.

TECHNOLOGY AND CURRENT USES

It is difficult to imagine modern society without aluminum products. Most of the following uses of aluminum are familiar to daily life and activities.

Aluminum products are common around the house. Aluminum metal is used in the manufacture of a wide variety of cans, cooking

Aluminum powder is used as a pigment and for its vigorous reactivity in rocket fuels and fireworks, like sparklers. *(Yellow/Shutterstock)*

utensils, and food packages. Aluminum foil is a common kitchen item. Home construction materials—doors, siding, windows, screens, and rain gutters—are made of aluminum.

Because of its light weight, aluminum finds many applications in transportation and in aerospace. Aluminum is used extensively in the manufacture of aircraft frames, engines, landing gear, and cabins. Motor vehicle materials—radiators, engine blocks, transmissions, and body panels—can all be made of aluminum. Rapid transit vehicles—light rail trains, for example—often are constructed of aluminum. Locomotives, freight cars, and cargo containers may be made of aluminum. Boat hulls may be constructed of aluminum. Also, to save weight, satellites are made of aluminum. Aluminum's light weight is also useful in sporting equipment. Skis, baseball bats, and tennis rackets are primarily made of aluminum.

Manufacturers often take advantage of aluminum's corrosion resistance. Highway signs are almost exclusively made of aluminum. Electrical wire and cable may be made of aluminum. Aluminum often is used for coating mirrors because of its high optical reflectivity.

Aluminum powder is used as a pigment in paints and for its vigorous reactivity in rocket fuels, explosives, and fireworks. In particulate form, aluminum oxide is used as an abrasive in emery board and sandpaper. Aluminum chloride and other aluminum compounds are common in antiperspirants.

2

Gallium

Gallium—element 31 in the periodic table—is a hard, brittle, silvery gray metal with a density of 5.92 g/cm^3. It ranks 34th in abundance in Earth's crust, with two naturally occurring isotopes—$^{69}_{31}Ga$ and $^{71}_{31}Ga$. Unlike most liquids, which contract in volume as they begin to solidify, gallium is an unusual element in that the liquid expands as it begins to solidify (just as water expands as it freezes).

With the exceptions of mercury and cesium, gallium has the lowest melting point (86°F [30°C]) of all metals. In addition, gallium's high boiling point (3,999°F [2,204°C]) gives gallium the largest temperature range of any element over which it is in the liquid state at atmospheric pressure. With such a large liquid range, gallium would make an ideal liquid for thermometers if not for its reactivity with glass at high temperatures. Otherwise, gallium could readily replace mercury in thermometers used to measure high temperatures.

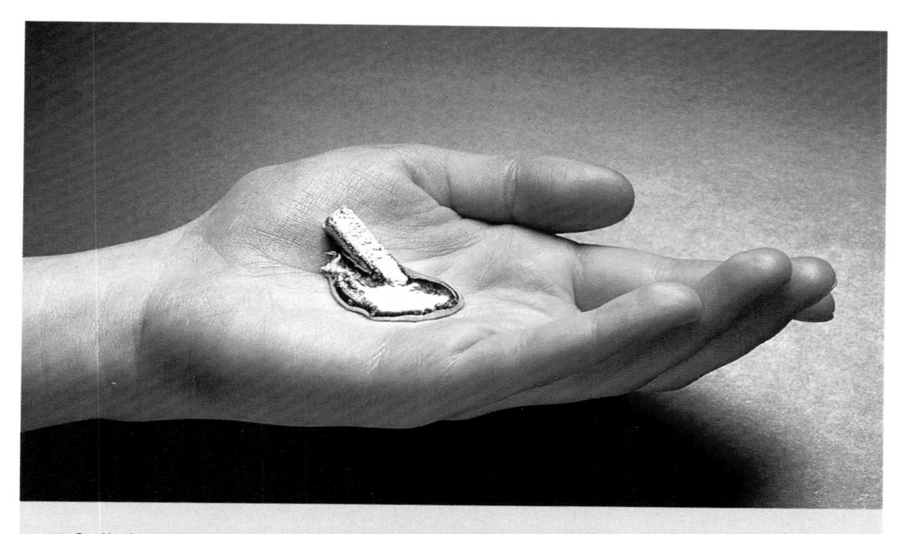

Of all the elements, gallium has the largest temperature range over which it is in the liquid state at atmospheric pressure. *(Charles D. Winters/Photo Researchers)*

There are no predominantly gallium ores; gallium is found only in trace amounts in ores of other metals like aluminum and zinc. About 95 percent of the gallium consumed goes into electronic devices such as transistors. Gallium has an electronic configuration of [Ar] $4s^2 3d^{10} 4p^1$, giving gallium three valence electrons, like the other elements in group IIIA. Therefore, gallium's only important ion is the "+3" ion. It is in this form, either as gallium arsenide (GaAs) or as gallium phosphide (GaP), that gallium is used in semiconductors.

THE ASTROPHYSICS OF GALLIUM

Gallium nuclei are not built in the cores of average stars like the Sun. Neither are they made in supernova explosions—as are many elements heavier than iron. Instead, gallium (like selenium) builds up over thousands of years in the atmospheres of more massive stars via neutron capture and electron release, with the requirement that iron 56 be available as a "seed." Neutrons become available from the capture of alpha particles by ^{13}C and ^{22}Ne as follows:

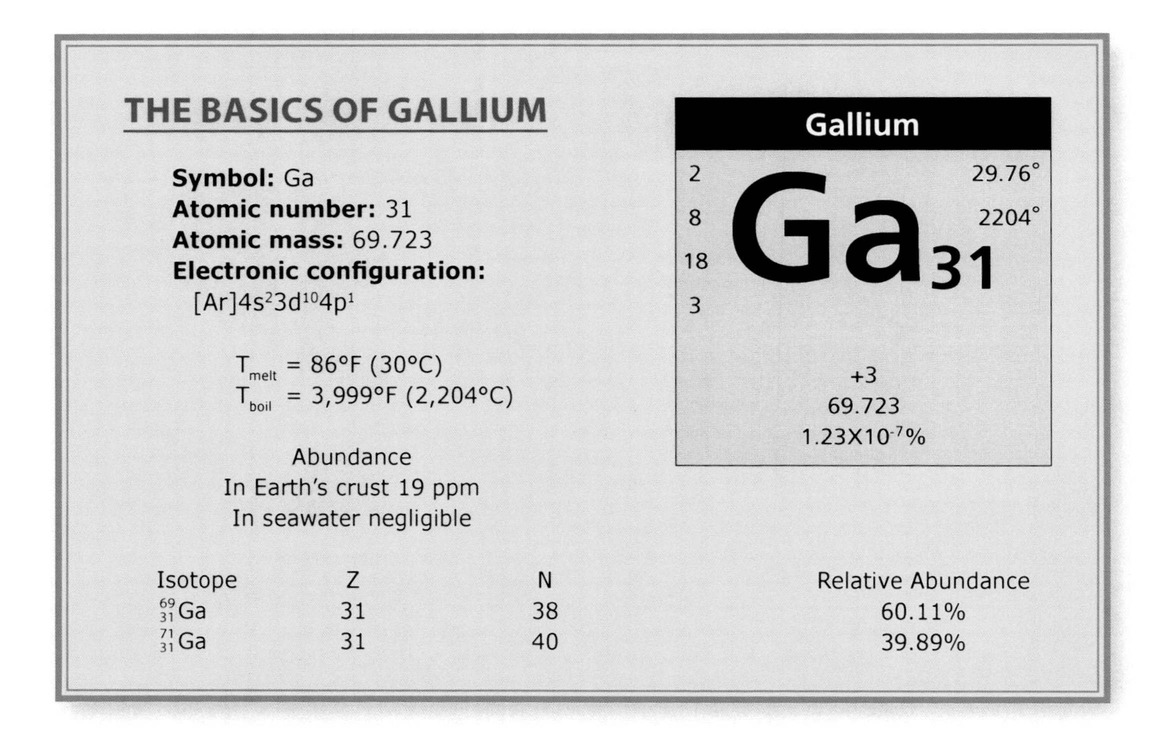

$$^{13}_{6}C + ^{4}_{2}\alpha \rightarrow ^{16}_{8}O + ^{1}_{0}n;$$

$$^{22}_{10}Ne + ^{4}_{2}\alpha \rightarrow ^{25}_{12}Mg + ^{1}_{0}n.$$

Because the synthesis proceeds slowly due to a low density of neutrons, it is called the slow process or "s-process." Gallium is formed when a stable zinc 68 nucleus acquires an additional neutron. The zinc 69 nucleus thus formed is unstable and decays as follows:

$$^{69}_{30}Zn \rightarrow ^{69}_{31}Ga + e^{-} + \bar{\nu}.$$

The *antineutrino* $\bar{\nu}$ ensures conservation of energy in the collision. The gallium thus produced then flows into the surrounding atmosphere on the stellar wind and mixes with the interstellar medium (ISM). As stars like the Sun form, they collect the surrounding gases and incorporate a little gallium into their makeup.

A puzzling phenomenon is an observed gallium excess over the solar abundance in so-called chemically peculiar (CP) stars. Astronomers distinguish CP stars by their strange characteristics, such as high magnetic fields, variable spectra, slow rotation, and changes in brightness. The gallium enhancement is a subject of ongoing scientific discussion and research. Because these stars rotate slowly, mixing may be lessened, allowing layering of different elemental species. This can occur when gravitational effects dominate over radiation pressure, which could produce more gallium near the surface of the star. This would enhance the apparent, but not the actual, abundance. Another hypothesis proposes that high magnetic fields associated with these stars allows the accretion of extra gallium from the surrounding atmosphere, which would boost the actual abundance in the star. The answer to the gallium question will not be resolved without further investigation by astrophysicists.

DISCOVERY AND NAMING OF GALLIUM

Mendeleev's version of a periodic table of the elements is particularly useful, even today, because chemists are able to predict the existence and approximate properties of undiscovered elements. Mendeleev arranged the elements in rows in order of increasing atomic weight so that elements that exhibit similar physical and chemical properties would also fall into the same column. When the next known element in terms of atomic weight would end up in the "wrong" column, Mendeleev left one or more blank spaces so that the next known element could be placed in a column containing other elements with similar properties.

After zinc, arsenic was the next heaviest element known at that time. Placing arsenic immediately following zinc would have placed arsenic in the same column as aluminum. Arsenic, however, is much more like phosphorus than it is like aluminum. Therefore, Mendeleev left a blank space in the table after zinc. In fact, Mendeleev left two blank spaces—one below aluminum and the other below silicon—so that arsenic would appear under phosphorus. He claimed that the two spaces he skipped corresponded to elements that had not yet been discovered, elements that he tentatively named *eka-aluminum* and *eka-silicon*. He even predicted what some of the elements' properties should be—including their atomic weight, density, melting and boiling points, and formulas for their compounds with oxygen and chlorine. Mendeleev lived to see his predictions come true as the discoveries of gallium and germanium filled in his blank spaces.

Gallium's discoverer was Paul-Emile Lecoq de Boisbaudran (1838–1912), a French chemist. Boisbaudran set out deliberately to try to find the element missing in the periodic table between aluminum and indium. Suspecting that the unknown element might be present in the mineral sphalerite (zinc sulfide, ZnS), in 1874, Boisbaudran began to investigate the composition of more than 100 pounds of ore containing sphalerite. Boisbaudran was well grounded in the use of atomic spectra as an analytical tool. Consequently, in 1875, when he observed two spectral lines that did not correspond to lines of any known element, he suspected that he was detecting the presence of a new element. The quantity he had obtained of the new element, however, was too small to isolate a sample.

Boisbaudran repeated the experiment using a much larger sample of ore—several hundred pounds of it. In addition to zinc, the ore was found to contain traces of many other elements, including copper, arsenic, lead, aluminum, iron, cobalt, chromium, cadmium, indium, thallium, mercury, selenium, silver, bismuth, tin, and antimony. Fortunately, the ore contained enough gallium that he was able to isolate a 1-gram sample. To honor his native country, Boisbaudran chose to name his new element *gallium,* after the name *Gaul,* which was what the land in France was known as during the Roman Empire. It was Mendeleev who, when he heard of Boisbaudran's discovery, quickly identified gallium as *eka-aluminum.*

In 1876, Boisbaudran and an assistant started with an even larger sample of the ore—almost 9,000 pounds—and succeeded in isolating 75 grams of gallium. Such a large sample of gallium permitted the determination of its atomic weight, density, melting point, volatility, reactivity with acids, and compounds formed by combination with oxygen and chlorine. All of these properties were found to be in close agreement with Mendeleev's predictions. In fact, one of Mendeleev's predictions had even been that eka-aluminum would most likely be discovered by spectroscopic means, which it was!

Gallium was found later also to be present in deposits of aluminum ore, feldspar, mica, and basalt.

THE CHEMISTRY OF GALLIUM

Even at high temperatures, gallium resists oxidation in air. Gallium, however, does dissolve readily in both strong and weak acids to form the gallic ion (Ga^{3+}). Because both aluminum and gallium compounds

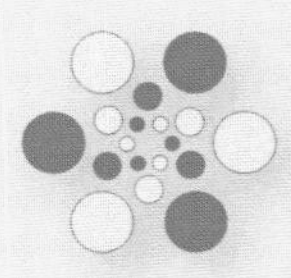

GALLIUM IN SEMICONDUCTORS

Perfectly conducting materials allow electrons to move freely among the atoms of the material. At normal temperatures, there are no perfect conductors, but metals like silver and copper are very good.

Semiconductors, on the other hand, rely on the jumping of electrons between accessible energy levels for current to flow in the material. In a solid, these levels are more numerous than in a gas of the same species and form broad bands of available energies, allowing for large congregations of electrons. These electrons undergo random thermal motion, but cannot jump to the next energy band if the gap between bands is much greater than the average thermal energy. So current flow in semiconductors is strongly dependent on the ambient temperature. Electrons can also be made to jump between bands if they absorb energy in any other form, such as electromagnetic energy. An applied electric field or light absorption can make current flow in semiconductors. It is this property that makes them useful in circuits—small amounts of current can be made to flow at prescribed moments. This is an important property in transistors, *optoelectronics*, and solar cells.

The energy bandwidths and gaps are also dependent on the exact nature of the material. Silicon and germanium, for example, are good semiconductors, while carbon is not, though all three have similar electronic structure. (Note that they inhabit the same column in the periodic table.) The difference is that, in solid form, the gaps between the two highest energy levels in silicon and germanium are much smaller than in carbon, so electrons are able to transition between them with ease. The addition of tiny bits of impurities can also have an impressive effect on the bandwidths and gaps.

A small amount of arsenic added to crystal gallium has resulted in a material (gallium arsenide or GaAs) that is particularly well adapted to many important semiconductor applications. Compared to the more common silicon semiconductors,

(continues)

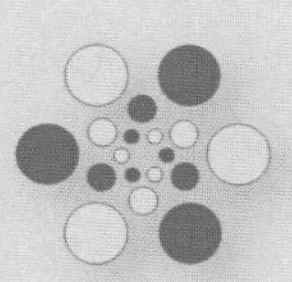

(continued)

GaAs displays relatively low sensitivity to heat, higher electron mobility, and the ability to withstand higher applied voltages. This means that GaAs transistors can be operated at higher frequencies and higher power with less associated noise. The nature of the band gap in GaAs also means that it is better at emitting light than silicon semiconductors, making it a material of choice for light-emitting diodes (LEDs). Though gallium is much more expensive than silicon, its high absorptivity allows for very thin films to be used as efficient collectors for solar cells. It has been used for this purpose in several Mars rovers.

Challenges for the GaAs applications include its expense. Researchers are working on finding appropriate inexpensive substrates for growth of the GaAs wafers or films. In addition, arsenic is toxic and carcinogenic, so great care must be taken toward worker safety in the wafer-polishing process.

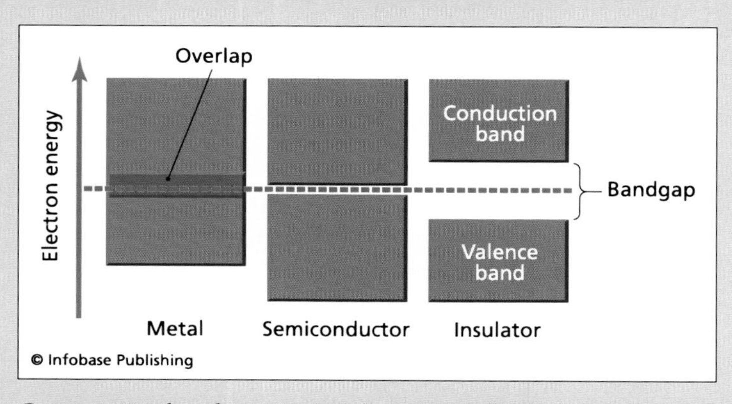

Comparative bandgaps in metals, semiconductors, and insulators

predominantly contain the metal in the "+3" state, the chemistry of gallium closely resembles the chemistry of aluminum. In analogy to the compounds of aluminum, gallium hydroxide ($Ga(OH)_3$) is insoluble, but other common compounds—$Ga_2(SO_4)_3$, $Ga(NO_3)_3$, and $GaCl_3$—are all soluble in water.

Compounds containing the Ga^+ and Ga^{2+} ions have been shown in the laboratory to exist, but they are unstable and have no commercial value.

Few practical uses of gallium compounds were known until the development of semiconductor technology. Because of the importance of gallium arsenide in semiconductors, GaAs is the most commercially important gallium compound.

The simplest qualitative test for gallium is by spectral analysis. The presence of gallium in a mixture can be detected by gallium's spectral lines at wavelengths of 417.2 and 403.3 nm.

THE GALLEX EXPERIMENT

Neutrinos (symbolized as ν)—nearly massless particles that accompany most nuclear fusion reactions—have presented challenges to physicists for decades. One such challenge has historically been called "The Solar Neutrino Problem." According to the standard model, which unifies three of the four known forces in nature, the Sun should produce a particular number of neutrinos each second, but experiments designed to count solar neutrinos consistently came up short. Scientists reasoned that the problem was with the detectors: They could not see low-energy neutrinos. A new type of detector was needed.

In 1991, a giant vat (100 tons [90.8 metric tons] or 14,000 gallons [52,996 L]) of gallium trichloride in solution with hydrochloric acid became the definitive solar neutrino laboratory—the gallium experiment, or GALLEX—deep inside a mountain in central Italy. The reaction of interest was as follows:

$$\nu + {}^{71}Ga \rightarrow {}^{71}Ge + e^-,$$

which is triggered only by low-energy neutrinos of the sort produced in the primordial proton-proton chain that produces helium in stars. From 1991 to 1997, GALLEX counted the frequency with which a neutrino reacts with gallium, an extremely uncommon occurrence, which is why the vat had to be so enormous. Six years of operation obtained only 80 reactions per second for each 10^{36} target atoms—80 solar neutrino units (SNU).

From 1991 to 1997, the GALLEX experiment counted the frequency with which a neutrino reacts with gallium. *(INFN-Gran Sasso National Laboratories)*

Curiously, this was still well below the number predicted by theory, which could not explain a result below 132 SNU. The key turns out to be neutrino type. In the above reaction equation, the neutrino is of the particular type associated with electrons. There are two other types, muon neutrinos and tau neutrinos, both generally associated with reactions concerning their partner particles—*muons* and *tau particles*. Experiments that were nearly simultaneous with GALLEX have shown that neutrinos can change type. This very recent understanding made particle theorists change course because it means neutrinos must have mass, an idea that had been put forth but largely ignored owing to a lack of experimental evidence. The rigor of GALLEX may well have been the trigger that made particle theorists decide to investigate other, less likely possibilities.

TECHNOLOGY AND CURRENT USES

Almost all of the world's gallium supply is used in the electronics industry along with other applications and may be found in radios, televisions, and calculators. Gallium arsenide and gallium phosphide are used in a variety of semiconductor devices, including diodes and laser diodes, and may be found in computers and cell phones. Gallium arsenide is also used in solar batteries. Small amounts of gallium are found in thermometers, solders, arc lamps, batteries, and catalysts.

3

Indium and Thallium

Indium is element 49; it is a silvery gray metal with a density of 7.31 g/cm^3. Indium is a very soft metal—softer than lead—and is *malleable* and *ductile.* A malleable substance is one that can be hammered easily into different shapes. A ductile substance is one that can be drawn into a wire.

Thallium is element 81; it is a silvery white metal with a density of 11.85 g/cm^3. Only trace amounts of either metal are found in Earth's crust, and there are no significant ores of either one. Essentially no thallium is obtained from ores within the United States; most thallium is imported from Belgium, Canada, and the United Kingdom.

Indium most often occurs as an impurity in zinc ores, while thallium occurs as an impurity in copper, zinc, and lead ores and in pyrite minerals (a group of metal sulfides that includes "fool's gold"). There are

Thallium is a silvery white metal; most thallium is imported from Belgium, Canada, and the United Kingdom. *(Tomihahndorf)*

practical applications for both. Thallium is extremely toxic, so a common use in its history has been as a poison.

THE ASTROPHYSICS OF INDIUM AND THALLIUM

Indium and thallium, both being heavier than iron, are synthesized in supernovae via the rapid capture by iron nuclei of a succession of neutrons, which is called the "r-process." Some fraction of these nuclei also

Indium most often occurs as an impurity in zinc ores. *(Russell Lappa/Photo Researchers, Inc.)*

THE BASICS OF INDIUM

Symbol: In
Atomic number: 49
Atomic mass: 114.818
Electronic configuration:
$[Kr]5s^24d^{10}5p^1$

T_{melt} = 314°F (157°C)
T_{boil} = 3,762°C (2,072°C)

Abundance
In Earth's crust 0.16 ppm

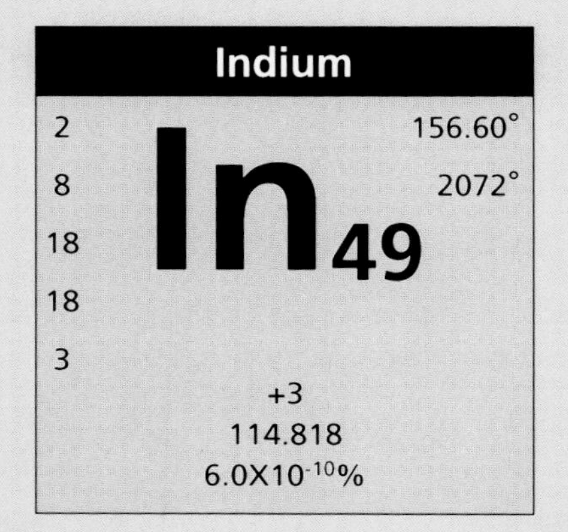

Isotope	Z	N	Relative Abundance
$^{113}_{49}In$	49	64	4.29%
$^{115}_{49}In$	49	66	95.71%

THE BASICS OF THALLIUM

Symbol: Tl
Atomic number: 81
Atomic mass: [204.382; 204.385]
Electronic configuration:
$[Xe]6s^24f^{14}5d^{10}6p^1$

T_{melt} = 579°F (304°C)
T_{boil} = 2,683°F (1,473°C)

Abundance
In Earth's crust 0.53 ppm

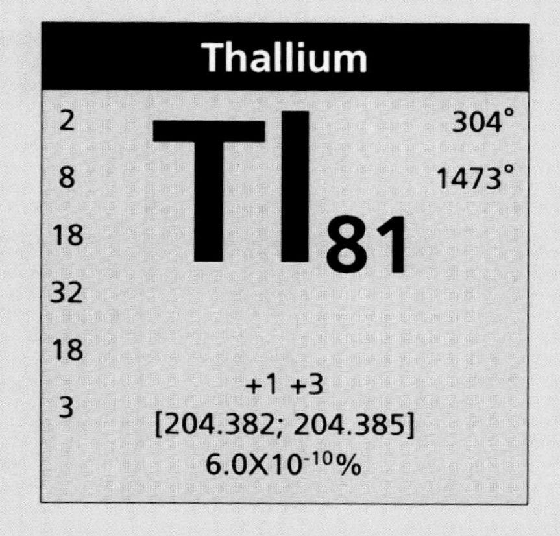

Isotope	Z	N	Relative Abundance
$^{203}_{81}Tl$	81	122	29.52%
$^{205}_{81}Tl$	81	124	70.48%

builds up slowly over thousands of years in the atmospheres of large-mass stars via neutron capture and electron release, with the requirement that iron 56 nuclei—remnants of prior supernova explosions—be available as seeds. Because this synthesis proceeds relatively slowly due to a low density of neutrons, it is called the "s-process."

Recent analysis indicates that indium and many other heavy elements can additionally be formed in supernovae via proton capture, or the "p-process." Supernova explosions produce a high density of photons that collide with some of the nuclei (in the photon- or "γ-process"), breaking them up and freeing protons that are subsequently captured by other nuclei to make higher-Z elements. The universal nature of this process is uncertain, however. Further research in this area is considered crucial to an understanding of galactic chemical evolution.

The solar abundance of indium has for many years been a curiosity to astrophysicists because it appears to be about six times higher than the observed abundance in meteorites. Since meteorites presumably formed during the collapse of the solar nebula, the mismatch is puzzling, especially since the thallium content is not nearly as depleted. The thallium and indium abundance in meteorites should be similar because they have similar condensation temperatures. Recent careful analysis of the solar spectrum, however, hints at discrepancies in past analyses and a probable mistaken identity of the spectral line attributed to indium.

Another puzzling phenomenon is an observed thallium excess in chemically peculiar (CP) stars, where spectra indicate an enhancement four *orders of magnitude* (10,000 times) over the solar abundance. (See "The Astrophysics of Gallium," in chapter 2, for a description of CP properties.) This thallium excess is a subject of ongoing scientific discussion and research.

DISCOVERY AND NAMING OF INDIUM AND THALLIUM

Credit for the discovery of indium goes to Ferdinand Reich (1799–1882) and his laboratory assistant Hieronymus Richter (1824–98), both of Germany. Reich was a professor of physics and inspector at the Freiberg School of Mines. He was a particularly talented teacher who was well liked by his students. Ferdinand Reich made several other important contributions to science. Having studied chemistry in Paris, he introduced the French metric system of weights and measures to his colleagues in Saxony. In addition, Reich determined the average density of Earth to be about 5.48 g/cm^3. In England, Henry Cavendish made the same determination. Their two values were in close agreement with the modern value of 5.52 g/cm^3.

Richter was a chemist at the School of Mines and eventually became the director. Although indium is found in small amounts in only a few ores, its discovery was relatively straightforward. Indium imparts a brilliant indigo color to the flame of a Bunsen burner. Reich was colorblind, so he enlisted Richter to make the spectroscopic observations. In 1863, Reich and Richter first observed an intense indigo line in the *spectrum* of a sample of zinc ore. This line did not correspond to a spectral line of any other known element. Recognizing that they had discovered a new element, Reich and Richter named it *indium* after its indigo line. They were able to prepare small, impure samples for analysis.

Thallium was discovered almost simultaneously by two independent scientists: Sir William Crookes (1832–1919) in England and Claude-Auguste Lamy (1820–78) in France. Crookes was professor of physics and chemistry at the Royal College of Chemistry in London. Lamy was president of the French Society of Chemistry. In 1861, Crookes observed a brilliant green line in the spectrum of selenium that did not correspond to a line of any known element. Crookes called the new element *thallium* because of its green spectral line (*thallus* is Latin for a green plant shoot). Crookes was elected a fellow of the Royal Society for his discovery of thallium. Subsequently, many honors were bestowed upon him, including the Royal Medal, the Davy Medal, the Copley Medal, knighthood, membership in the Order of Merit, and election to the presidency of the Royal Society.

Unaware of Crookes's discovery, Lamy also isolated thallium in 1861. In fact, in 1862, both men exhibited samples of metallic thallium at the International Exhibition in London. Since both of them claimed to be thallium's discoverer, this led to a feud over priority of discovery. The feud was settled by giving both men equal credit. Crookes probably observed thallium's green line first, but Lamy isolated the first sample from a mixture of pyrites. In addition, Lamy studied thallium's chemistry and demonstrated its extreme toxicity. Thallium was found to be an anomaly among metals. Although its "+1" ion has chemical properties similar to those of the alkali metals, its "+3" ion has properties similar to those of aluminum.

THE CHEMISTRY OF INDIUM AND THALLIUM

Indium is a relatively rare metal because it concentrates in only a few ores—mainly ores of zinc and lead. The metal is easily obtained by reduction with zinc metal, or by electrolysis. Because of indium's scarcity and softness, it has few uses. Its chemistry is governed by a general trend that occurs among families of elements: Upon descending a column, there is a general tendency for the atoms to increase in size. The first element in Group IIIA is boron, which consists of small atoms and is a metalloid that tends to form covalent compounds instead of ionic compounds. Just below boron are aluminum and gallium, the atoms of which are larger than the atoms of boron and the chemistry of which tends to involve both covalent and ionic bonding in the "+3" oxidation state. The naturally occurring element in column IIIA with the largest atoms is thallium, at the bottom of the column. As the sizes of the atoms increase, the chemistry shifts from domination by the "+3" oxidation state to the "+1" state. By the time thallium is reached, its chemistry is dominated by ions with a "+1" charge. The chemistry of indium—whose atoms are intermediate in size between aluminum and gallium above it, and thallium below it—reflects both kinds of chemistry. Indium's chemistry includes both "+1" and "+3" ions, sometimes in the same compound, which is unusual for an element in Group IIIA. The "+1" ion In^+ is the *indous* ion, and the "+3" ion In^{3+} is the *indic* ion.

Because of indium's rarity and relatively few uses, its chemistry is not as well studied as the chemistry of the more common elements in the periodic table. There is even some ambiguity about the exact chemistry of indium. In addition, many of indium's compounds are unstable in aqueous solution. Given those qualifications, the following table exhibits some of the more common compounds of the Group IIIA elements and shows the trend in going from mostly "+3" oxidation states to mostly "+1" oxidation states upon descending the column. In the

ELEMENT	+1 COMPOUNDS	+3 COMPOUNDS	BOTH +1 AND +3 COMPOUNDS
Boron	None	BCl_3 (as B_2Cl_6 dimer)	None
Aluminum	None	$AlCl_3$, Al_2O_3, AlN	None
Gallium	Uncommon	Ga_2Cl_6, Ga_2O_3, GaN, GaAs	None
Indium	InCl	$InCl_3$, In_2O_3, InN	In_2Cl_4 (i.e., $InCl\cdot InCl_3$)
Thallium	TlCl, Tl_2O	Uncommon	None

vapor state, the halides of indium form *dimers* in which two molecules link together. For example, indium chloride ($InCl_3$) becomes In_2Cl_6.

In addition, indium forms stable compounds containing a "+2" ion. This ion may be analogous to the fairly common mercurous ion, Hg_2^{2+}, and would have the formula In_2^{2+}.

In aqueous solution, compounds of the thallous ion (Tl^+) dominate thallium chemistry. Thallium's chemistry tends to be similar to Group I elements like silver and the alkali metals. Because of this similarity, thallium is often found in association with ores of those elements. Like indium, pure thallium metal is obtained either by reduction with zinc metal or by electrolysis.

In analogy to the halides of the silver ion (AgCl, AgBr, and AgI), halides of the thallous ion (TlCl, TlBr, and TlI) also are insoluble in water. Likewise, in analogy to the ions of the alkali metals—Na^+ and K^+, for example—most other compounds of Tl^+ are soluble in water.

The presence of indium or thallium in mixtures is best detected using spectral analysis. Indium has a characteristic indigo line, and thallium has a characteristic green line.

FROM SEMICONDUCTING TO SUPERCONDUCTING

In semiconductors, performance properties depend on the exact nature of the material. As discussed in chapter 2, GaAs semiconductors display relatively low sensitivity to heat, have high electron mobility, and are able to withstand high applied voltages. It turns out that the addition of indium to GaAs allows for the fabrication of transistors that can be operated at extremely high frequencies (above 100 GHz) with very low associated noise. Crystals of InGaAs are grown on an indium phosphide (InP) substrate and are, therefore, often called InP semiconductors. As with any material developed for a specific purpose, there are negative aspects. For one, indium of the required purity is expensive and becoming less available due to the closure of several mines. Also, the addition of indium to electronic components significantly reduces the useful operating voltage, so the applied power must be decreased. And indium is very brittle. Nevertheless, InP semiconductors will be useful in the future, with particular promise in cell-phone technology.

Thallium, too, has its applications in semiconductor technology, particularly as the compound thallium oxide, which has good transparency and high electrical conductivity, and is easily grown as a thin film. Another semiconductor compound—thallium bromide—is proving useful for the detection of X-rays and gamma rays.

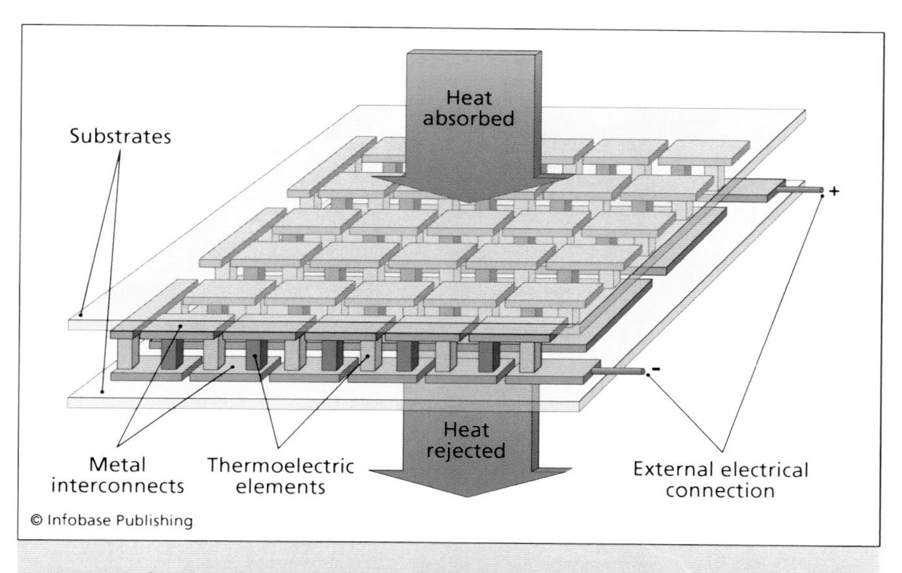

A typical semiconductor thermoelectric module. If a temperature difference is maintained across the module, electrical power will be delivered to an external load and the device will operate as a generator.

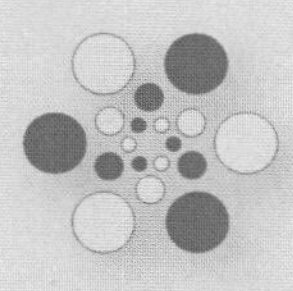

THALLIUM: A LEGENDARY POISON

Thallium is toxic, and thallium compounds are absorbed readily through the skin. Like lead and mercury, it is a cumulative poison because the body has no mechanism for excreting it. Also like lead and mercury, it can cause deadly nervous system disorders. Because of its chemical similarity to sodium and potassium, thallium tends to travel through the body together with those ions. Historically, thallium poisoning has been related to the use and misuse of rodent and ant poisons. While banned long ago in the United States, some countries still allow the use of thallium-sulfate rodenticides. Since thallium is available so readily, it has frequently been used for the purpose of homicide. A particular symptom of thallium poisoning is loss of hair, a pivotal clue found in mystery writer Agatha Christie's 1961 novel *The Pale Horse.*

An antidote for thallium poisoning is a nontoxic pigment called Prussian blue. Potassium in Prussian blue replaces thallium in the body by a mechanism called *ion exchange,* in which potassium ions replace thallium ions. As Prussian blue passes through the circulatory system, the thallium that has been absorbed is flushed from the body.

Thallium poisoning plays a featured role in Agatha Christie's 1961 mystery novel, *The Pale Horse. (Tobi Zausner)*

Perhaps the most intriguing research on thallium is in the field of high-temperature superconductivity, where it has been shown that a lead-tellurium semiconductor becomes superconducting when doped with a small thallium impurity. This thallium-doped lead-telluride may turn out to be a crucial breakthrough for transferring heat energy to electrical (and therefore usable) energy at typical automobile engine temperatures.

TECHNOLOGY AND CURRENT USES

Although less use is made of indium than of other elements near indium in the periodic table, there are some applications in industry. For example, indium is used to solder wires to germanium transistors and is a component of germanium transistors. In analogy to gallium, indium arsenide, indium antimonide, and indium phosphide are used in semiconductors.

When police officers administer breath analyses for possible DUI ("driving under the influence") violations, indium compounds are used to preserve the breath samples.

Indium is used in corrosion-resistant bearings. Some indium is mixed with palladium in dental alloys.

Indium alloys tend to have very low freezing points, making them useful in switches. In addition, indium, in combination with silver and cadmium, has been used in control rods for nuclear reactors.

Because of its toxicity, there are few uses for thallium. However, some use is made of thallium in photo cells and other optical materials. Glass that contains thallium can provide a protective coating on semiconductors and capacitors.

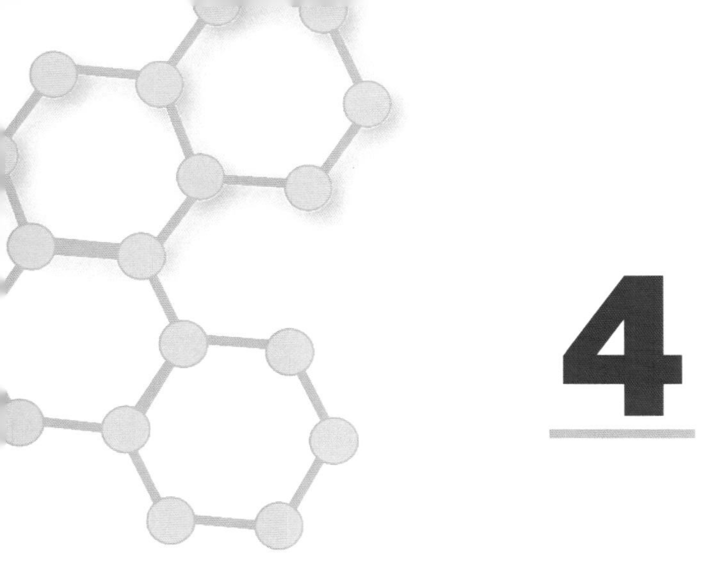

4

Tin

Tin is element number 50; it is a silvery gray metal with a density of 7.31 g/cm^3. Although it is only the 49th element in abundance, tin deposits tend to be well concentrated. Tin is one of the elements used extensively by ancient people. More than 4,000 years ago, people learned the technique of alloying tin with copper to make bronze, which allowed the manufacture of weapons and other implements that were stronger than pure copper alone.

Metallic tin and its salts are nontoxic. Thus, an important use of tin is for coating the insides of steel food containers. Tin is more resistant than steel is to attack by acids common in foods. Without a thin coating of tin, the steel would corrode easily, adversely affecting both the quality and taste of the food inside.

The most important mineral that contains tin is cassiterite (SnO_2). Only 16 percent of tin used in the United States comes from domestic

The Chinese fashioned weapons of bronze for thousands of years. *(The Art Archive/Hubei Provincial Museum/Laurie Platt Winfrey)*

sources. Mostly, tin is mined in a number of other countries around the world, including Malaysia, Indonesia, Thailand, Brazil, Bolivia, Zaire, Australia, Nigeria, China, and England.

Like all metals, tin only forms positively charged ions. Both the stannous ion (Sn^{2+}) and the stannic ion (Sn^{4+}) are common.

THE ASTROPHYSICS OF TIN

Like gallium, tin nuclei build up over thousands of years, typically in the atmospheres of asymptotic giant branch (AGB) stars, via neutron

THE BASICS OF TIN

Symbol: Sn
Atomic number: 50
Atomic mass: 118.710
Electronic configuration:
$[Kr]5s^2 4d^{10} 5p^2$

T_{melt} = 449°F (232°C)
T_{boil} = 4,716°F (2,602°C)

Abundance
In Earth's crust 2.2 ppm

Tin
2 8 18 18 4
Sn 50
231.93°
2602°
+2 +4
118.710
$1.25 \times 10^{-8}\%$

Isotope	Z	N	Relative Abundance
$^{112}_{50}Sn$	50	62	0.97%
$^{114}_{50}Sn$	50	64	0.66%
$^{115}_{50}Sn$	50	65	0.34%
$^{116}_{50}Sn$	50	66	14.54%
$^{117}_{50}Sn$	50	67	7.68%
$^{118}_{50}Sn$	50	68	24.22%
$^{119}_{50}Sn$	50	69	8.59%
$^{120}_{50}Sn$	50	70	32.58%
$^{122}_{50}Sn$	50	72	4.63%
$^{124}_{50}Sn$	50	74	5.79%

capture and electron release, with the requirement that iron 56 be available as an initial "seed." Because the synthesis proceeds slowly due to a low density of neutrons, it is called the slow process or "s-process." (See chapter 2.) Tin is thus formed when a nucleus of indium 115, which is synthesized via the s-process but is unstable, decays as follows:

$$^{115}_{49}In \rightarrow {}^{115}_{50}Sn + e^- + \bar{\nu}.$$

Tin is also synthesized in recurring stellar explosions called *nova* outbursts, which generate a high density of protons, allowing heavy nuclei to form by a series of proton captures.

Tin is one of the few elements for which a relatively large data sample exists regarding its interstellar abundance, which appears to be 20–30

A high abundance of tin has been detected in interstellar dust. *(The Art Archive/Hubei Provincial Museum/Laurie Platt Winfrey)*

percent higher than the solar abundance. If so, it is the only known element with a higher abundance in the interstellar medium than in the Sun. Analyses of emission lines from singly ionized tin atoms seem to indicate an ongoing transfer of tin ions between the gas and dust states in diffuse interstellar clouds. Further investigation will be needed to fully understand the processes involved.

DISCOVERY AND NAMING OF TIN

Tin has been used since ancient times. In Mesopotamia, by about 3500 B.C.E., bronze—an alloy consisting of about 80 percent copper and 20 percent tin (not to be confused with *brass,* an alloy of copper and zinc)—was already in use. In fact, the discovery of how to make bronze marked the transition from the Stone Age to the Bronze Age. Because bronze is harder and more corrosion-resistant than pure copper, bronze was used to manufacture tools, ornaments, statues, bells, cups, urns, vases, battle-axes, helmets, knives, shields, and swords. In the Middle East, between about 1500 and 1000 B.C.E., iron replaced bronze in many applications, marking the beginning of the Iron Age. There are no set dates, however, that demarcate the beginning or the end of the Bronze Age. Different regions of the world experienced their Bronze Ages at different times. In addition, there were considerable overlaps in time in the beginning, when implements associated with the Bronze Age were used together with implements associated with the Stone Age, and toward the end, when Bronze Age implements were used together with Iron Age implements.

DOUBLY MAGIC

Tin 100 and tin 132 are known as doubly *magic* isotopes. In nuclear physics, the term *magic* refers to nuclei that have closed energy shells. These are analogous to filled electron orbits in the Bohr model of the atom; when one orbital level reaches its capacity, the particles cannot force their way in, but instead begin building a new grouping. So it is with nuclei. Protons and neutrons band together in energetically efficient ways. It turns out that groupings of 2, 8, 20, 50, 28, 82, and 126 make filled nuclear shells. As is the case with filled-shell atoms, filled-shell nuclei are particularly stable. Noble gases do not readily give up electrons. Magic isotopes do not readily give up nuclei.

The shells fill in magic numbers of protons and neutrons separately. A nucleus with a magic number of protons is pretty stable. A nucleus with a magic number of neutrons is pretty stable. A nucleus with both a magic proton number and a magic neutron number is really stable (compared with its neighboring isotopes). These nuclei are, therefore, doubly magic.

These notions have been discussed for perhaps a century, but only in the last decade have scientists been able to make heavy, doubly magic isotopes. Only a handful of accelerator laboratories worldwide are capable of the production of tin 100 and 132. Tin 132 with 50 protons and 82 neutrons is of great interest because it is unusually neutron rich, which borders on an untenable state. There comes a point where a nucleus cannot keep an excess of neutrons. This state is not well understood and exists for a number of elements in the vicinity of tin on the periodic table. Tin 132, however, has a half-life 40 times longer than its isotopic neighbors.

Physicists continue to explore the doubly magic regime because of the need to understand neutron nucleosynthesis as it applies to astrophysics. Theory and experiment have not yet merged. Heavy-nuclei research continues to be an exciting area of exploration.

THE CHEMISTRY OF TIN

Tin belongs to an exceptional family of elements—Group IVA. Starting at the top of the column, carbon is a nonmetal. Descending the column, silicon and germanium are metalloids. At the bottom of the column, tin and lead are metals. Since nonmetals, metalloids, and metals have such distinctly different chemical and physical properties, there are no *group* properties of which to speak. Effectively, tin and lead are more similar to the IVB transition metals (titanium, zirconium, and hafnium) than to the other IVA elements.

Tin and lead are both fairly soft metals with low melting points. Both metals are sufficiently available in workable deposits, and both have been utilized since antiquity in a number of applications. Both elements form "+2" and "+4" ions. Therefore, much of what may be said about tin applies to lead also.

Metallic tin is easily obtained by reduction of the ore with carbon, which is the reason tin metallurgy was well known in ancient times: The reduction of tin could be accomplished in wood fires without requiring more modern forms of technology. Tin has three distinct crystalline structures in the solid state. The most malleable form is called *white tin,* which is the form that is rolled into very thin sheets called *tinfoil.* Tinfoil resists corrosion. Therefore, it is very commonly used in the food-canning industry to line the insides of steel cans, where it protects the steel from corrosion by the acids found in food.

Tin forms several useful alloys with other metals. Bronze is an alloy of tin and copper. *Solder* is an alloy of tin and lead. *Babbitt* is an alloy of tin, antimony, and lead that formerly was used for ball bearings. *Pewter* is almost pure tin with a small amount of copper. Pewter is used to make

Tin is quite malleable and was used extensively during the late 19th and early 20th centuries to make decorative ceiling panels. *(Steven Belanger/Shutterstock)*

household items like candlesticks, plates, bowls, tableware, and vases. Some of the finest examples of pewter date from American Colonial times. (After 1800, chinaware was used more often because chinaware was less expensive than pewter.) *Tinplate* is the alloy of tin and iron that is used in the inner lining of cans.

Both stannous ions (Sn^{2+}) and stannic ions (Sn^{4+}) readily form a number of simple compounds. Examples of their compounds are shown in the following table:

ION	COMMON COMPOUNDS
Stannous, Sn^{2+}	$SnCl_2$, $Sn(NO_3)_2$, SnO, $Sn(OH)_2$, SnS, $SnSO_4$
Stannic, Sn^{4+}	$SnCl_4$, $Sn(NO_3)_4$ SnO_2, $Sn(OH)_4$, SnS_2, SnH_4

The chlorides and nitrates are soluble in aqueous solution, while most other tin compounds are insoluble. Stannic sulfide (SnS_2) has long been used to gild objects with a yellow color. Because of its resemblance to gold, it is referred to as *mosaic gold*.

Stannous sulfide (SnS), which is brown in color, and stannic sulfide (SnS_2), which is yellow in color, form when hydrogen sulfide gas (H_2S) is bubbled through solutions of Sn^{2+} or Sn^{4+} salts. These reactions are illustrated in the following equations:

$$Sn^{2+} (aq) + H_2S (g) \rightarrow SnS (s) + 2 H^+ (aq);$$

$$Sn^{4+} (aq) + 2 H_2S (g) \rightarrow SnS_2 (s) + 4 H^+ (aq).$$

Stannic sulfide is amphoteric and will redissolve in high concentrations of sulfide ions to form a thiostannate ion, as shown by the following reaction:

$$SnS_2 (s) + S^{2-} (aq) \rightarrow SnS_3^{2-} (aq).$$

Metallic tin dissolves in hydrochloric acid to form the Sn^{2+} ion, as shown by the following reaction:

$$Sn (s) + 2 HCl (aq) \rightarrow SnCl_2 (aq) + H_2 (g).$$

In hot nitric acid, tin dissolves to form the Sn^{4+} ion according to the following reaction:

$$3 Sn (s) + 16 HNO_3 (aq) \rightarrow 3 Sn(NO_3)_4 (aq) + 4 NO (g) + 8 H_2O (l).$$

In the "+2" oxidation state, tin can form the stannite ion, SnO_3^{4-}. In the "+4" oxidation state, tin can form the stannate ion, SnO_3^{2-}. Combined with sodium or potassium ions, Na_2SnO_3 or K_2SnO_3 can be formed.

The qualitative detection of Sn^{2+} or Sn^{4+} in mixtures makes use of the ease with which these ions are reduced to tin metal. Adding a few drops of mercuric chloride ($HgCl_2$) will cause a gray mixture of mercurous chloride (Hg_2Cl_2, which is a white solid) and metallic mercury (which is gray) to form, which is the definitive indicator of the original presence of one or both of the tin ions. The reactions that occur are shown in the following equations:

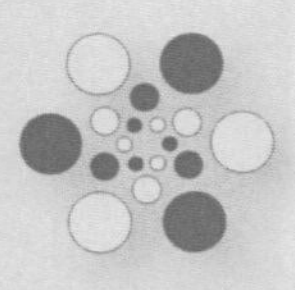

MAKING BRONZE

Sometime around 3000 B.C.E., metal workers began to realize that copper with small impurities was preferable to pure copper for making tools. This was probably an accidental discovery, since most natural copper deposits contain impurities—often as tin or arsenic. Bronze is typically now defined as copper with a 5–15 percent tin impurity. (Bronze and *brass* should not be confused. Bronze is an alloy of copper and tin; brass is an alloy of copper and zinc). One advantage is that bronze has a significantly lower melting temperature (950°C [1,742°F]) than copper (T_{melt} = 1,084°C [1,983°F]), so not as much fuel was needed for heating the furnaces. In addition, hammered bronze is much harder than hammered copper and holds an edge better. In the early Bronze Age, smiths also used arsenic bronze, but it was phased out, probably owing to a recognition of the health hazards. (See chapter 8.) Smiths who worked with arsenic had long-term exposures to its toxic fumes, and many developed crippling nerve damage in their limbs.

A huge tin trade developed in Assyria (now Iraq), Anatolia (now Turkey), Britain, Spain, Germany, and China. Tin mines in Assyria alone shipped out about a *tonne* per year between 1950 and 1850 B.C.E. Bronze seemed the ideal material for a multitude

$$Sn\ (s) + 2\ HgCl_2\ (aq) \rightarrow SnCl_2\ (aq) + Hg_2Cl_2\ (s);$$

$$SnCl_2\ (aq) + Hg_2Cl_2\ (s) \rightarrow SnCl_4\ (aq) + 2\ Hg\ (l).$$

The exact appearance depends on the relative quantities of tin and mercuric chloride that are mixed together.

TECHNOLOGY AND CURRENT USES

Because of tin's abundance, resistance to corrosion, and nontoxic nature, tin finds a large number of applications in modern society. For example, tin is used to coat other metals to protect them from corro-

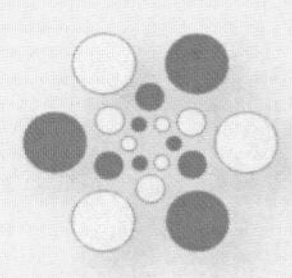

of uses, from cooking to farming to warfare. It is an exceptionally strong material. A thin layer of copper oxide tends to form on bronze with exposure to weather, but once that layer is formed, the underlying bronze remains intact, which is why many ancient bronze artifacts are still in good condition today.

Around 1500–1000 B.C.E., however, bronze manufacture declined, and iron and its alloys began to gain popularity, even though iron has a higher melting point (1,538°C [2,800°F]). The Bronze Age evolved slowly into the Iron Age, largely due to the depletion of forests in close proximity to tin mines. By 1200 B.C.E., wood harvesting to fuel the kilns had led to massive deforestation in much of the eastern Mediterranean. The Rio Tinto mines in Spain used up to 287 tons (260 tonnes) of wood per day at the height of production. The system of making bronze at or near the metal's source—the mines—became unsustainable. And while tin could be shipped, some sources were mined out, and trade routes suffered periodic disruption.

While bronze is still used today in bells and cymbals, sculptures, and various technologies that require springs and bearings, iron (and steel, which is iron with impurities) has become the dominant industrial metal worldwide.

sion and is used in various kinds of solders. It is used as a coating for food and beverage containers, signs, toys, and kitchen housewares. Some hardware items and automotive parts are made of tinplate, as are some coatings for gasoline tanks, roofing materials, and radiator water tubes.

Tin has applications in the aircraft industry and may be used in electronic applications, such as plating printed circuit boards. Niobium-tin (Nb_3Sn) is a superconductor that functions well even in the presence of high magnetic fields. Nanoscale structures made of tin have exhibited unconventional and intriguing melting behavior.

Small amounts of tin exist naturally in most fruits and vegetables, though it is not clear how much is needed for health. There is some evidence that supplementation may be beneficial to people who suffer from fatigue, headaches, or asthma. Stannous fluoride (SnF_2) has been used as a source of fluoride in toothpaste.

5

Lead and Bismuth

Lead, element 82, and bismuth, element 83, are the heaviest elements in the periodic table that have stable isotopes. Beginning with polonium—element 84—any element heavier than lead or bismuth has only radioactive isotopes. Lead is a soft, bluish white solid with a relatively high density of 11.34 g/cm^3. Bismuth is a reddish white solid with a density of 9.78 g/cm^3.

Lead has been known and used since ancient times—as early as 4000 B.C.E. It is never found as the free metal. When reduced to the metal, however, it is very soft. On the *Moh's hardness scale* that geologists use, lead has a hardness of only 1.5 (where 1, talc, is the softest and 10, diamond, is the hardest). Lead's principal minerals are galena (PbS), crocite ($PbCrO_4$), anglesite ($PbSO_4$), massicot (PbO), and cerrusite ($PbCO_3$). Nearly all of the lead used in the United States—92 percent—is mined in the United States. The remaining 8 percent comes

Galena (lead sulfide) is the most common mineral containing lead. *(Jens Mayer/Shutterstock)*

from Canada, Australia, Mexico, and Peru, with some mining also occurring in Russia, Germany, Morocco, and Spain.

Bismuth can occur as the free metal. Its principal minerals are bismuthinite (Bi_2S_3) and bismutite ($(BiO)_2CO_3$), found mostly in South

Very pure bismuth can form astonishing crystals when it cools from its molten state. *(Larry Stepanowicz/Visuals Unlimited)*

THE BASICS OF LEAD

Symbol: Pb
Atomic number: 82
Atomic mass: 207.2
Electronic configuration:
[Xe]$6s^2 4f^{14} 5d^{10} 6p^2$

T_{melt} = 621°F (327°C)
T_{boil} = 3,180°F (1,749°C)

Abundance
In Earth's crust 10 ppm

Lead

2	327.46°
8	1749°
18	Pb_{82}
32	
18	+2 +4
4	207.2
	$1.03X10^{-8}$%

Isotope	Z	N	Relative Abundance
$^{204}_{82}Pb$	82	122	1.48%
$^{206}_{82}Pb$	82	124	23.6%
$^{207}_{82}Pb$	82	125	22.6%
$^{208}_{82}Pb$	82	126	52.3%

More than 25 other unstable lead isotopes exist, with a total relative abundance of 0.02 percent.

THE BASICS OF BISMUTH

Symbol: Bi
Atomic number: 83
Atomic mass: 208.98040
Electronic configuration:
[Xe]$6s^2 4f^{14} 5d^{10} 6p^3$

T_{melt} = 520°F (271°C)
T_{boil} = 2,847°F (1,564°C)

Abundance
In Earth's crust 0.025 ppm

Bismuth

2	271.40°
8	1564°
18	Bi_{83}
32	
18	+3 +5
5	208.98040
	$4.7X10^{-10}$%

Isotope	Z	N	Relative Abundance
$^{209}_{83}Bi$	83	126	100%

Traces of 17 unstable bismuth isotopes have been observed, most with short half-lives.

America. Although bismuth has medicinal uses, lead is toxic. Lead products must be used in ways that do not expose people to the possibility of ingesting lead compounds or breathing lead dust.

THE ASTROPHYSICS OF LEAD AND BISMUTH

Lead and bismuth—like gallium and tin—are mostly produced by slow-neutron capture in massive stars, particularly asymptotic giant branch (AGB) stars. Unlike the alpha-capture process, these "s-process" events allow for a steady buildup along the isotopic chain of each element, eventually forming closed or "filled" neutron shells. As described in chapter 4, such filled neutron shells are particularly stable. A nucleus that has a filled neutron and proton shell is termed *doubly magic*. The heaviest known doubly magic isotope is lead 208, which has 82 protons and 126 neutrons. Though capable of producing many elements heavier than iron, the s-process is not energetic enough to form elements heavier than bismuth.

The abundances of lead and bismuth in a certain class of massive stars that may have formed very soon after the big bang—the so-called Population III stars—is a current topic of intrigue among astrophysicists. Such stars, if they existed, could explain the ubiquity of heavy elements in the universe. The supernova stage of a Population III star would have occurred very early in the evolution of the universe, spewing elements as heavy as iron into the surrounding space. Some scientists, however, hypothesize that these stars could have had unusual convection properties that allowed a partial mixing of protons, which, in turn, could have led to an extensive stellar layer where slow-neutron-capture synthesized elements as heavy as lead and bismuth. The search continues for astrophysical evidence of Population III stars, which would further scientific understanding of the very early universe.

Closer to home, but no less curious, is the recent discovery of lead and bismuth "snow" on the surface of Venus. While this classic greenhouse planet is much too warm for Earthlike snow, at high Venus altitudes, conditions seem to be cool enough for lead- and bismuth-sulfide to condense out of the surrounding atmosphere and fall to the ground. This phenomenon was first observed in 1995 as bright, reflective areas in mountainous regions of the planet. A lander mission to Venus that could collect soil samples would be able to validate the hypothesis and even use radioactive dating (see the following section) to calculate the age of Venus.

THE RADIOACTIVE NATURE OF LEAD AND BISMUTH

Lead has four stable isotopes—^{204}Pb, ^{206}Pb, ^{207}Pb, and ^{208}Pb. The last three (notated in the rest of this section as $^{20\times}$Pb) are the final products of the radioactive decay chains of uranium and thorium beneath the Earth's surface. (See decay chain diagram.) Because the number of $^{20\times}$Pb nuclei increases in this manner, while ^{204}Pb abundance remains constant, isotopic variations can be used to determine the relative ages of various rock samples.

The $^{20\times}$Pb/^{204}Pb ratios are important to geologists who study Earth's continental and oceanic evolution. Measurements of lead isotopic ratios help geologists understand sequences of mineralizing events, such as the distribution of gold deposits in the eastern United States and the underlying root rock that preserves a distinct geologic period.

Several different radioactive bismuth isotopes are also produced in the uranium and thorium decay chains. Bismuth 210, in particular, is proving useful in studies of volcanism. Its five-day half-life and tendency to take on the gas phase make this isotope fairly easy to monitor along with ^{210}Pb in volcanic aerosols. This provides a method for studying the abundance of the U-238 parent in inaccessible material deep underground that feeds volcanoes.

DISCOVERY AND NAMING OF LEAD AND BISMUTH

Lead is a relatively heavy, soft, malleable, and toxic metal. The ancient Egyptians were already working with lead in 3500 B.C.E. The Chinese had produced lead by 3000 B.C.E. The Phoenicians and Romans mined lead in Spain. Lead pipes were used in plumbing systems beginning in early Babylonia in the third century B.C.E. and continued from the time of the Roman Empire to modern times. The Romans used lead in plumbing systems, facial powders, mascaras, paints, food seasonings, coinage, and wine (as a preservative). In addition, the Romans fashioned dinner plates, goblets, pitchers, and cookware from lead. Personal uses of lead, however, resulted in lead poisoning, which historians believe

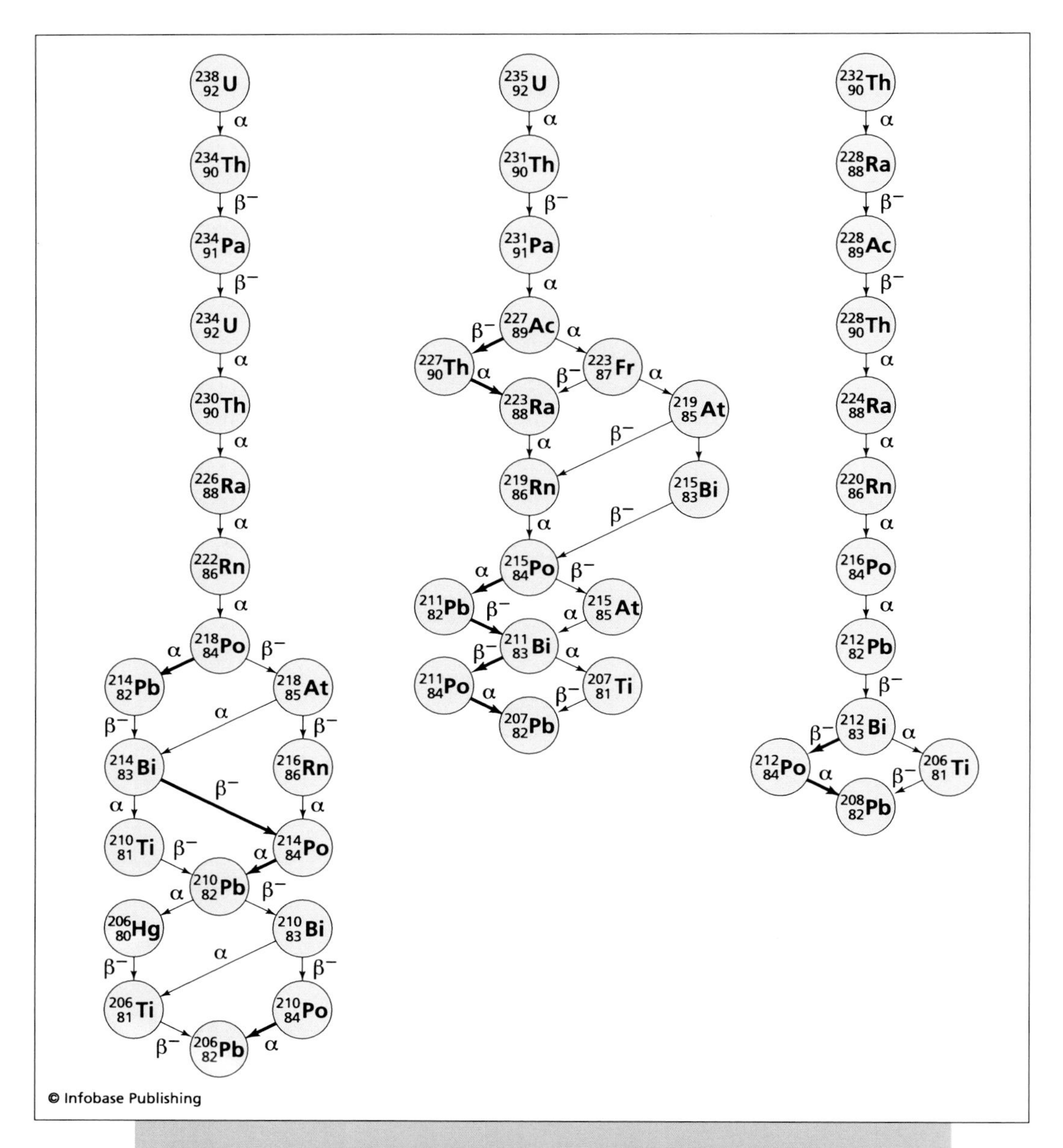

Decay chains of ^{238}U, ^{235}U and ^{232}Th: β^- indicates an electron is released in the decay, whereas α indicates a helium nucleus is released in the decay.

contributed to the ultimate decline and fall of the Roman Empire. In plumbing systems today, older buildings may still have lead drainpipes, but most plumbing is now done using aluminum, copper, or plastic.

Alchemists considered lead to be the fundamental metal. During the Middle Ages, numerous efforts were made to try to *transmute* lead into gold. Their efforts were completely unsuccessful, because no ordinary chemical reaction can convert any element into another element. A nuclear reaction is required, and medieval alchemists had no knowledge of nuclear reactions.

The term *plumbing* and the chemical symbol for lead, Pb, are derived from the Latin name for lead, *plumbum*. The name *lead* is related to the old German word *lot*, which meant "weight" or "plummet."

Medieval alchemists referred to bismuth as early as the 15th century, but there is no record of bismuth's discovery, initial isolation, or recognition as an element. Even bismuth's name gives no clue as to its origin. *Bismuth* is simply a corruption of the original German name, *wismut*, which is probably a compound formed from the German words for white *(weiss)* and mass *(masse)*.

Bismuth has been employed in a variety of uses from at least 1400 C.E. to the present. Thin layers of metallic bismuth were used to cover boxes, caskets, chests, and cupboards. Shortly after the invention of the printing press, print type began to be cut from a bismuth alloy. At that time, however, it was unclear whether bismuth was a separate element, or a different form of another element such as antimony, tin, zinc, or lead. It was not until 1753 C.E. that bismuth was identified as a unique element by the French nobleman Claude-François Geoffroy.

THE CHEMISTRY OF LEAD AND BISMUTH

As stated in chapter 4, the chemistry of lead is very similar to the chemistry of tin. Like tin, lead forms two ions—the plumbous ion (Pb^{2+}) and the plumbic ion (Pb^{4+}). These ions form a number of compounds that are completely analogous to the compounds formed by the stannous and stannic ions, as shown in the following table:

ION	COMMON COMPOUNDS
Plumbous, Pb^{2+}	$PbCl_2$, PbO, $Pb(OH)_2$, PbS, $PbSO_4$
Plumbic, Pb^{4+}	$PbCl_4$, PbO_2, $Pb(OH)_4$, PbS_2

Bismuth was identified as a unique element by the French nobleman Claude-François Geoffroy in 1753. *(Academie des Sciences archive)*

Another important lead compound is lead chromate ($PbCrO_4$), which has a bright yellow color. For many years, lead chromate was an important pigment in paint, as were lead sulfate ($PbSO_4$) and lead carbonate ($PbCO_3$), which are white in color. In recent decades, however, lead in paints has been phased out because of lead's toxicity. A common substitute for lead in white paint is titanium dioxide (TiO_2), which is completely nontoxic.

A common source of lead is the mineral galena (PbS). Galena dissolves in hydrochloric acid to give hydrogen sulfide gas (H_2S) and an aqueous solution of lead chloride ($PbCl_2$), as shown in the following equation:

$$PbS\ (s) + 2\ HCl\ (aq) \rightarrow PbCl_2\ (aq) + H_2S\ (g).$$

Hydrogen sulfide is a toxic gas, but it can be captured and converted into sulfuric acid (H_2SO_4) for industrial and commercial applications. Sulfide ores of a number of metals are, in fact, the major source of sulfur for manufacturing sulfuric acid. The lead chloride that is produced is fairly easily reduced to the metal.

Lead has a relatively low melting point that lends itself to an application familiar to hunters—its use to make bullets. Unfortunately, just as lead is toxic to humans, it is also toxic to wildlife. Many animals and birds (ducks and geese, for example) survive gunshot wounds only to die later, at which time their remains are consumed by a variety of *carrion* eaters and *scavengers*—such as vultures, ravens, bald eagles, bears, and coyotes. Ingesting lead in this manner significantly increases their *mortality*. Concern about mortality among other birds and animals—especially species on the Endangered Species List, like the California condor—has resulted in programs conducted by game and fish departments to substitute steel ammunition (which is nontoxic) for lead. As an incentive to use nontoxic ammunition, steel bullets are sometimes distributed for free in hunting areas.

Hydrochloric acid oxidizes metallic lead to the "+2" state, as shown in the following equation:

$$Pb\ (s) + 2\ HCl\ (aq) \rightarrow PbCl_2\ (aq) + H_2\ (g).$$

Lead monoxide (PbO) can be oxidized by sodium hypochlorite (NaClO) to lead dioxide (PbO_2) according to the following reaction:

$$PbO\ (s) + ClO^-\ (aq) \rightarrow PbO_2\ (s) + Cl^-.$$

PbO is orange-yellow in color. PbO_2 is white. So-called *red lead* is a mixture of the two oxides and has a formula of Pb_3O_4.

In the VA family of elements, bismuth is the heaviest member that occurs naturally. Upon descending the column, the elements begin with nitrogen and phosphorus, which are nonmetals. They are followed by arsenic and antimony, which are metalloids, and finally by bismuth, which is a metal. Thus, bismuth's properties are similar in some ways to the other elements in its family, yet it has some unique properties as the only metal in the family.

Many bismuth alloys expand upon heating, but a number of bismuth alloys also melt at relatively low temperatures. Therefore, these latter alloys are useful in applications such as temporary plugs, as they

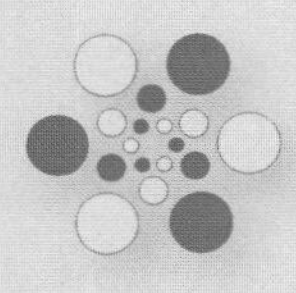

THE COOLING OF REACTORS

The energy produced by nuclear reactors comes from the fission of atoms, which results in nuclear fragments moving at high velocities and, therefore, extreme heating of the core material. The core container is surrounded by a coolant, which absorbs the heat and transfers it to a steam generator. Early reactors used water as the coolant of choice, but selected molten metals are better because they have higher boiling points. Sodium-based metal coolants have been used, but these combust upon contact with both air and water (clearly a hazardous situation), so they must be isolated by an intermediate coolant loop, which adds expense.

Lead and lead-bismuth coolants, however, are nearly ideal for use in reactors. They are largely nonreactive with air and water, are fairly inexpensive, and have low vapor pressures at typical reactor operating temperatures. Cooling is by natural convection, so such reactors are inherently safe against coolant loss. In addition, lead is a natural radiation shield.

Lead-bismuth, whose melting temperature is 124°C, is preferable to pure lead (T_{melt} = 327°C) because lead coolant could solidify at low operating temperatures. Bismuth atoms in the coolant, however, can capture neutrons that escape from the core via the following reaction:

$$^{209}_{83}\text{Bi} + ^{1}_{0}\text{n} \rightarrow \gamma + ^{210}_{83}\text{Bi}.$$

Bismuth 210 then decays to polonium 210—an alpha emitter and potential health hazard to workers.

Another difficulty with these materials is that lead and lead-bismuth coolants are highly corrosive to steel. Other materials are being studied for lining the containment vessels, and Russian researchers have had some success in the use of vanadium. It is also helpful to minimize the oxygen content in the coolant.

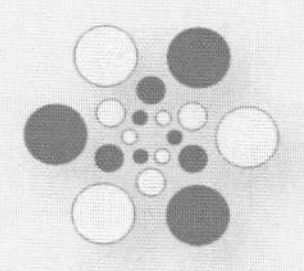

Molten lead-bismuth is the coolant of choice for the so-called Generation IV reactors, which are in development at the time of this writing. These would feature completely enclosed pre-manufactured modules with uranium alloy fuel cores surrounded by an outer shell containing the coolant. At the end of a typical useful lifetime of about 15 years, the module would be allowed to cool, and could then be shipped as a self-shielded, self-contained unit. Inherently safe, relatively compact, and economically competitive, such reactors will be marketed for electricity production on small grids and to developing countries.

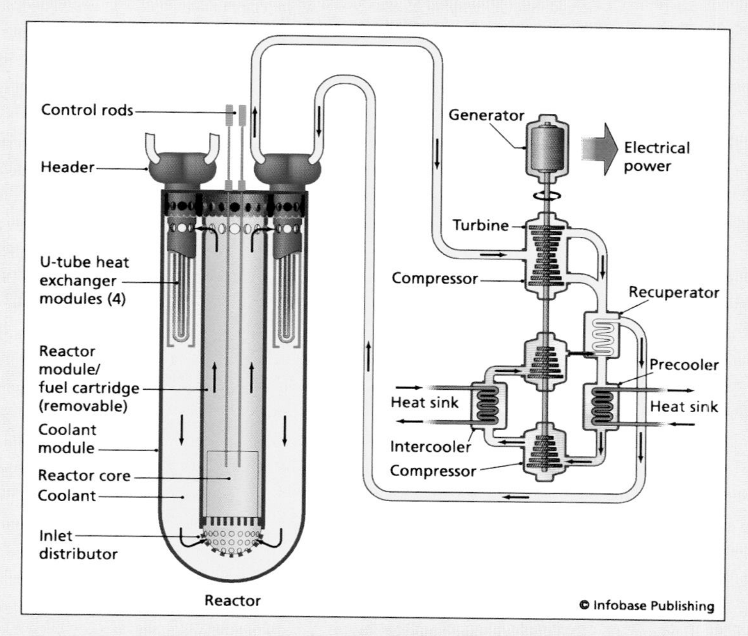

Lead-bismuth eutectic (Pb-Bi) is superior to sodium as a coolant for fast reactors.

melt and expand when heated. Examples of these applications include water sprinklers and automatic-closing fire doors.

As a group VA element, bismuth can exist in compounds in the "+5" oxidation state. However, its most common oxidation state is "+3." In the "+3" state, common bismuth compounds include bismuth trioxide (Bi_2O_3), bismuth sulfide (Bi_2S_3), bismuth chloride ($BiCl_3$), and bismuth nitrate ($Bi(NO_3)_3$). Bismuthine (BiH_3) can be made, but it decomposes rapidly at room temperature. Compounds in the "+5" state include bismuth pentoxide (Bi_2O_5) and sodium bismuthate ($NaBiO_3$). In acidic solution, the bismuthate ion is a very powerful oxidizing agent. One example of its oxidizing power is its ability to convert the manganous ion (Mn^{2+}) to the permanganate ion (MnO_4^-), in which manganese is in the "+7" oxidation state. This reaction is shown in the following equation:

$$2\ Mn^{2+}\ (aq) + 5\ BiO_3^-\ (aq) + 14\ H^+\ (aq) \rightarrow$$
$$2\ MnO_4^-\ (aq) + 5\ Bi^{3+}\ (aq) + 7\ H_2O\ (l).$$

The manganous ion is a pale pink color, and permanganate is a deep purple color, so the appearance of the purple color is evidence that a reaction has taken place.

THE HEALTH HAZARDS OF LEAD

Until about 1970, lead was a component in most paints because it improved durability, resistance to corrosion, and thermal adaptability. By 1950, the ill effects of lead absorption into the human system had become widely acknowledged, and many paint manufacturers adopted a policy that limited the lead content in indoor paint to less than 1 percent. In 1973, the legal upper limit became 0.5 percent, and in 1978—only five years later—it was further reduced by nearly an order of magnitude to 0.06 percent. This arguably drastic reduction indicates the severity of the problem.

Lead has no known practical use in the human body, but it presents many hazards. Intake may occur by inhalation (paint fumes or dust from sanding a surface, for example) or swallowing (possibly from food served in lead-painted dinnerware or from inadvertent hand-to-mouth ingestion). Exposure leads to especially severe problems for children, who may experience retarded growth, brain and nervous-system dam-

Peeling paint that contains lead is a health hazard if it is inhaled or ingested. *(Konstantin Sutyagin/Shutterstock)*

age, headaches, or hearing problems. Adults most commonly notice problems from continued (chronic) exposure. Effects include high blood pressure, nervous-system disorders, fertility problems and miscarriages, muscle and joint pain, and loss of memory. *Acute* exposure results in vomiting, diarrhea, and even coma.

While lead was outlawed in indoor paint (and automobile fuel) nearly 40 years ago, it remains an important environmental concern. It is still allowed in paint for industrial use in structures like bridges and ships, exposing workers to its effects. Older homes with flaking paint also release lead into the atmosphere. This is of particular concern in impoverished neighborhoods that cannot afford the necessary renovations, and families living in these older homes may inhale or ingest toxins. When old buildings are repainted, they are first scraped and sanded, releasing lead into the atmosphere and into the surrounding soil. Ideas about federal regulations regarding lead pollution have been introduced in Congress, most recently by Barack Obama and Hillary Clinton during their senatorial tenures. Legislation seems to be on hold, however. Unfortunately, the problem of lead pollution still is not widely understood. Efforts at education will be key to alleviating future health effects resulting from lead exposure.

THE LEAD STORAGE BATTERY

One of the most familiar uses of lead is in batteries for motor vehicles, which are usually referred to as lead storage batteries. Lead storage batteries use both metallic lead and lead dioxide, which makes them the major use for lead dioxide. When the battery is discharging, the metallic lead serves as the *anode,* or negative electrode for the battery (the electrode that gives off electrons). The lead dioxide serves as the *cathode,* or positive electrode (the electrode that takes in electrons). At the anode, lead is oxidized from the neutral metal to the Pb^{2+} ion, while at the cathode, lead is reduced from the "+4" oxidation state in PbO_2 to the Pb^{2+} ion. A fairly concentrated solution of sulfuric acid forms the *electrolyte,* or electrically conducting solution, in the battery, so that the lead ions that form in both halves of the reaction (the so-called *half reactions*) combine with sulfate ions (SO_4^{2-}) to form lead sulfate ($PbSO_4$). These two half reactions are shown in the following chemical equations:

$$\text{(anode) } Pb\ (s) + H_2SO_4\ (aq) \rightarrow PbSO_4\ (s) + 2\ H^+\ (aq) + 2\ e^-;$$

$$\text{(cathode) } PbO_2\ (s) + 2\ H^+\ (aq) + H_2SO_4\ (aq) + 2\ e^- \rightarrow PbSO_4\ (s) + 2\ H_2O\ (l).$$

When these equations are added together, the electrons cancel, which results in the overall reaction shown below that takes place when power is being drawn from a battery:

$$Pb\ (s) + PbO_2\ (s) + 2\ H_2SO_4\ (aq) \rightarrow 2\ PbSO_4\ (s) + 2\ H_2O\ (l).$$

The reverse reaction takes place when a battery is being recharged through the alternator (or generator) in the engine. Passing an electrical current through water also causes the *electrolysis* of water to occur, producing hydrogen gas (H_2) and oxygen gas (O_2), as shown in the following reaction:

$$2\ H_2O\ (l) \rightarrow 2\ H_2\ (g) + O_2\ (g).$$

Because the hydrogen that is produced is potentially explosive, a solid catalyst is present inside the battery that recombines hydrogen and oxygen to make water again.

One *cell* of the battery, which consists of a plate of Pb, a plate of PbO_2, and the sulfuric acid, generates an electrical potential of about 2 *volts* (V). A typical automobile battery consists of six cells connected in *series,* which means that the potentials add together to make a 12-V battery overall. Batteries can also be made with only three cells, which would produce 6 V, or with more than six cells, which would produce more than 12 V. Besides cars and trucks, lead storage batteries are also used in motorcycles and golf carts. Electric vehicles run entirely on batteries, which must be recharged at regular intervals.

Most common metals would be expected to dissolve in sulfuric acid, lead included. As mentioned previously, lead does dissolve in

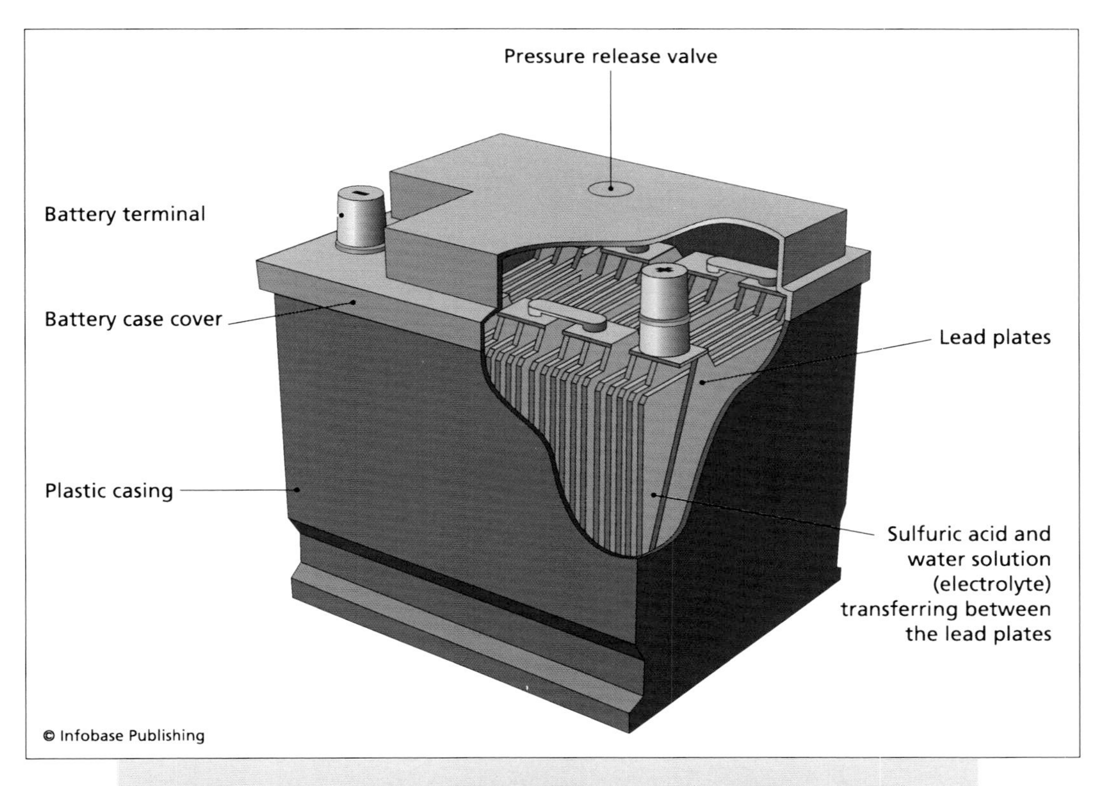

Schematic of a lead storage battery

hydrochloric acid. The reaction between metallic lead and concentrated sulfuric acid, however, is so extremely slow that it can be neglected for the lifetime of the typical automobile battery (5–6 years).

Lead storage batteries are extremely heavy for a reason. The electrical *current* that can be transmitted through the plates is directly proportional to the cross-sectional area of the plates. Since the main purpose of the battery is to start the engine, a very large current is required (about 100–200 *amperes* [amps, or A]), which in turn requires large plates. (In comparison, only a few amps are required to run accessories like the headlights or the radio.) Because the density of lead is high, the weight of the battery is also high.

TECHNOLOGY AND CURRENT USES

Lead has been used in plumbing for thousands of years, but it has recently been replaced in many applications with aluminum, copper, or plastic parts. It is currently used in storage batteries, especially for motor vehicles. Because consumers almost always trade in their old batteries when they buy new ones, a high percentage of lead from batteries is recycled. Historically, lead pigments were used in paints, but their use has been drastically reduced because of their toxicity. Lead is still used in industrial paint, some glass, and ammunition.

Because of its high density, lead is useful as a shield against X-rays and radioactivity. High-density lead weights are also used to balance tires on motor vehicles.

There are fewer uses of bismuth than there are of lead, but one medicinal use of bismuth is in the treatment of upset stomachs, a familiar over-the-counter example being Pepto-Bismol®. In addition, some cosmetics contain bismuth compounds. Bismuth alloys are also used in *thermocouple* devices, fire extinguishers, fire-door closers, and in lead-bismuth alloys for cooling nuclear reactors.

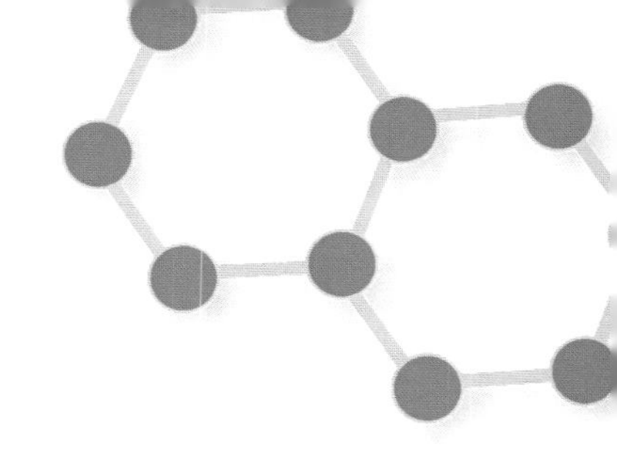

PART

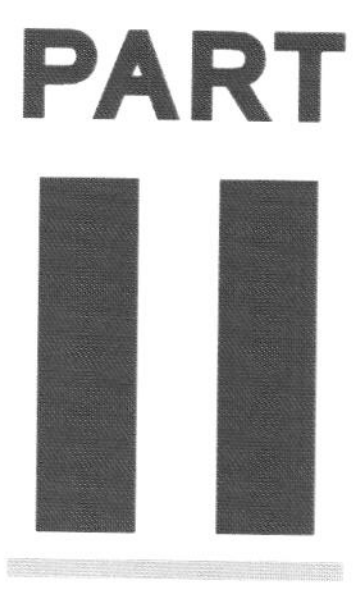

METALLOIDS

INTRODUCTION TO THE METALLOIDS

Most of the elements in the periodic table are metals. Metals tend to be located in the columns located on the left-hand side and middle of the table. Only 16 naturally occurring elements are classified as nonmetals. Nonmetals are located in the columns on the far right-hand side of the table. The metalloids are located in an intermediate region between the metals and nonmetals. There are only six naturally occurring elements that are unambiguously classified as metalloids: boron (B), silicon (Si), germanium (Ge), arsenic (As), antimony (Sb), and tellurium (Te). In this book, polonium (Po) also is being classified as a metalloid. Although astatine (At) sometimes is classified as a metalloid, it is treated in this set as one of the halogens.

The elements that most strongly exhibit properties associated with metals are those in the lower left-hand corner of the periodic table—cesium (Cs) being the primary example. As one progresses across a horizontal row *(period)* of the table from left to right, elements become less and less metallic in their characteristics. By the time elements on the right-hand side of the table have been reached, they are distinctly nonmetallic in nature. The metalloids represent that portion of the table in which the transition from metallic to nonmetallic characteristics is most apparent. Metalloids tend to exhibit most of the physical properties of metals, but the chemical properties of nonmetals.

The most important physical property of metals is their ability to conduct electricity. In this respect, metalloids (or *semimetals*) are *semiconductors* of electricity. Silicon and germanium are two of the most prominent metalloids, and play an important role in the semiconductor industry. Like metals, metalloids also tend to conduct heat, to be malleable and ductile, and to be shiny, or lustrous.

On the other hand, metalloids exhibit chemical properties that more closely resemble those of nonmetals. In compounds, metals only exist in positive oxidation states. Metalloids, however, like nonmetals, can exist in both positive and negative oxidation states. Metalloids can form compounds with both hydrogen and oxygen. Hydrogen will be in its usual +1 oxidation state, just as it is in compounds it forms with nonmetals. For example, silicon forms silane (SiH_4), in which silicon is in the "–4" oxidation state, in analogy to methane (CH_4), in which carbon is in the "–4" state. (In a neutral compound, the sum of the oxidation numbers has to equal zero: four atoms with "+1" and one atom with "–4" add to zero.) Similarly, oxygen will be in its usual "–2" oxidation state in combination with metalloids. Therefore, in silicon dioxide (SiO_2), silicon is in the "+4" state, in analogy to carbon dioxide (CO_2), in which carbon also is in the "+4" state. (One atom in the "+4" state and two atoms in the "–2" state add to zero.)

6

Boron

Boron is element number 5; it is a black, metallic-looking substance with a density of 2.46 g/cm^3. The principal source of boron is the mineral borax ($Na_2B_4O_7 \cdot 10H_2O$), with secondary sources being colemanite ($Ca_2B_6O_{11} \cdot 5H_2O$) and small quantities derived from volcanic spring waters. Boron is an essential nutrient for green algae and other plants and may have biological importance in animals.

The chemistry of most first-row elements is unique in comparison to the elements lying below them in the same family. Therefore, it can be misleading to compare boron too closely to aluminum. It is better to compare boron to silicon. In fact, boron oxide (B_2O_3) is extremely glasslike for an oxide, and, therefore, tends to be used in conjunction with silicon dioxide (SiO_2), the best-known glass-forming oxide.

THE BASICS OF BORON

Symbol: B
Atomic number: 5
Atomic mass: [10.806; 10.821]
Electronic configuration: $1s^2 2s^2 2p^1$

T_{melt} = 3,767°F (2,075°C)
T_{boil} = 7,232°F (4,000°C)

Relative Abundances
In Earth's crust 8.70 ppm
In seawater 4.44 ppm

Boron	
2	2075°
3	4000°
B_5	
+3	
[10.806; 10.821]	
$6.9X10^{-8}$%	

Isotope	Z	N	Relative Abundance
$^{10}_{5}B$	5	5	19.9%
$^{11}_{5}B$	5	6	80.1%

All boron ores are based on oxides of one composition or another. Tourmaline (a gem found in igneous rocks) is composed of a large number of elements, but is about 10 percent boron by weight.

THE ASTROPHYSICS OF BORON

Boron is one of the few elements that is not synthesized in stars. Until quite recently, its only known production pathway was the interaction of cosmic rays—mainly protons—with interstellar oxygen that floats around among the stars. This is a *spallation* reaction, which means a (fast) proton smashes into a (slow) oxygen nucleus with enough energy to break it up, resulting in interstellar ^{10}B and ^{11}B fragments.

It is difficult, however, to confirm that this is the predominant mode of boron production in space. One reason is that boron nuclei absorb and emit photons in the ultraviolet region of the electromagnetic spectrum, which is trickier to observe than the visible range. The few existing data indicate that there is something wrong with the spallation-only hypothesis. For example, measurements of the boron content in meteorites gives a value nearly twice what the theoretical model predicts. So researchers have been investigating other scenarios.

Studies of the famous *Supernova 1987a* led to a deeper understanding of astrophysical neutrinos and their role in nucleosynthesis. In Type II supernovae, neutrinos are produced in great numbers with extremely high energies. During the explosion, these high-energy neutrinos collide with carbon nuclei formerly made in the star's core and break them up into smaller nuclei like boron 11 (but not boron 10). This is now known as the "neutrino process" or "neutrino-induced spallation." In the same supernova explosion, small amounts of boron may also be produced by the capture of two alpha-particles onto helium 3 nuclei, producing boron 11.

DISCOVERY AND NAMING OF BORON

On Earth, sodium borate or "borax" has been used for thousands of years. The term *borax* comes from the Persian or Arabic word *borak* that means "white." Borax was used by ancient Egyptians when soldering gold and when practicing medicine and was brought to Europe by Marco Polo in the 13th century for Venetian goldsmiths to use.

Although borax and boric acid had been used for centuries, their chemical compositions were still a mystery as late as the beginning of the 19th century. It was not even known whether borax or boric acid occur naturally or if they can only be manufactured artificially. Investigation of the chemical composition of boric acid was able to occur in the early 1800s, following Humphrey Davy's successful isolation of sodium, potassium, magnesium, calcium, strontium, and barium. In 1808, Louis-Joseph Gay-Lussac (1778–1850) and Louis-Jacques Thénard (1777–1857) in France, and Humphrey Davy in England, independently reacted boric acid with metallic potassium, resulting in the isolation of a new element. Gay-Lussac and Thénard named the new element *bore;* Davy named it *boracium.*

Mineral exploration in Europe had yielded relatively small quantities of borax. Extensive deposits of borax were discovered in the 1860s in the Mojave Desert in southern California. There was no railroad to serve the mine, so ore was hauled by wagon teams of 20 mules. In the 1890s, the "20 MULE TEAM" symbol became the trademark of the Pacific Coast Borax Company, and the product today is still marketed as 20 Mule Team® Borax. In the early 1900s, borax and kernite

Borax is usually mined in open pits that can become quite extensive and unsightly. *(Mariusz S. Jurgielewicz/Feature Pix)*

($Na_2B_4O_7 \cdot 4H_2O$) were discovered in Kern County, California, where they are still mined using an open pit. The town of Boron, California, was named for boron (in contrast to the usual practice that a town's name exists first, and subsequently an element is named after it). Similar deposits were discovered in Death Valley, California (now a national park), and in the state of Nevada.

THE CHEMISTRY OF BORON

Boron is similar to carbon in its ability to build chains of molecules held together by covalent bonding. Boron is different from carbon, however, in that carbon atoms' four valence electrons favor the formation of straight- and branched-chain molecules and rings, whereas boron atoms' three valence electrons favor the formation of clusters of boron atoms. Boron has such great versatility that compounds of boron have been synthesized with virtually all of the other naturally occurring elements in the periodic table.

Boron hydrides were first prepared in analogy to carbon hydrides (CH_4, for example). Examples of boron hydrides include the following compounds: diborane (B_2H_6), tetraborane (B_4H_{10}), pentaborane (B_5H_9), and hexaborane (B_6H_{12}). These compounds have analogs in the hydrocarbons (C_2H_6, C_4H_{10}, C_5H_9, and C_6H_{12}). Whereas the hydrocarbons are all discrete molecules, however, the boranes tend to exist as dimers. B_2H_6, for example, consists of two BH_3 units linked together.

Boranes have a high-energy content and would make good fuels except that they tend to ignite at temperatures close to room temperature, making them very hazardous. Some use has been made of them, however, as additives in jet fuel (especially for high-altitude flying, where the pressure of oxygen is low). The principal characteristic of boron-containing compounds is that bonding tends to be largely covalent in nature. In other words, boron most commonly exists in the "+3" oxidation state in compounds, but the bonding is not ionic.

The story of the chemistry of boron is virtually synonymous with the life works of two distinguished chemists—Ukranian-American chemist Herbert C. Brown (1912–2004) and American chemist M. Frederick Hawthorne (b. 1928). Brown shared the 1979 Nobel Prize in chemistry with German chemist Georg Wittig (1897–1987). Brown began synthesizing boron hydrides as his doctoral work at the University of Chicago in the 1930s. When he began his work, the synthesis technique for diborane that was current at that time produced only three grams of diborane a day—too little for practical research. Brown discovered a new process for manufacturing diborane by reacting lithium hydride (LiH) with boron trifluoride (BF_3), as shown by the following equation:

$$6\ LiH\ (s) + 8\ BF_3\ (g) \rightarrow B_2H_6\ (g) + 6\ LiBF_4\ (s).$$

Once Brown had sufficient quantities of B_2H_6 with which to work, he succeeded in synthesizing numerous boron-containing compounds, including the following: sodium borohydride, $NaBH_4$; uranium (IV) borohydride, $U(BH_4)_4$; and sodium trimethoxyborohydride, $NaBH(OCH_3)_3$. (In uranium (IV) borohydride, the notation "IV" indicates that the uranium atom is in the "+4" oxidation state.)

Brown discovered that $NaBH_4$ acts as a mild *reducing agent* with respect to organic compounds. Its use became widespread in the syntheses of useful organic substances from *aldehydes, ketones,* and *acid chlorides.* In addition, he discovered that diborane reacts readily with *alkenes* as a means of producing *organoboranes* (compounds in which boron is attached to a hydrocarbon group). An example is illustrated by the following reaction of ethylene ($H_2C = CH_2$) with diborane:

$$H_2C = CH_2 + H - BH_2 \rightarrow H_3 - C - CH_2 - BH_2.$$

(Brown took personal pleasure in noting the sequence of letters in the main chain of the product: "H – C – B"—his own initials! He felt his parents had shown a lot of foresight when they named him.) Brown's active and productive career investigating the chemistry of boron spanned six decades. Organoboranes have a rich chemistry, the details of which are beyond the scope of this book. They have, however, numerous important applications in the chemical industry.

In March 2009, Fred Hawthorne was honored by the American Chemical Society for more than 50 years of research on boron chemistry. (Hawthorne was the recipient of the 2009 Priestley Medal, the highest award given by the American Chemical Society.) Like Brown, Hawthorne recognized that boron should exhibit chemical properties analogous to those of carbon, including the ability to combine with itself and form chains and rings of atoms. Examples of complex ions and compounds that were first synthesized or described by Hawthorne and his coworkers are $B_{12}H_{12}^{2-}$, $C_2B_{10}H_{12}$, and $Ni(C_2B_9H_{11})_2$. Compounds that Hawthorne has synthesized have found applications in catalysis, rocket fuels, and medicine. Some of Hawthorne's compounds can be used to deliver anticancer drugs to cancerous tumors. At present, Hawthorne is the director of the International Institute of Nano and Molecular Medicine at the University of Missouri in Columbia. An ongoing major field of study at the institute is boron neutron capture therapy (BNCT), a radiation treatment for diseases such as cancer and arthritis.

A HIGH-ENERGY FUEL

While hydrocarbons have been effective as fuels for automobiles and aircraft over the past several decades, the carbon dioxide by-product is

Boron hydride compounds have been investigated for use in high-powered rocket fuels. *(NASA [*Mariner 1 *Launch])*

a greenhouse gas that needs to be reduced in Earth's atmosphere, so the development of other fuels is a high priority. This is not the first time in recent history, however, that alternative fuels have been explored.

After World War II, when the United States and USSR were attempting to develop more powerful rocket engines and to extend the range of *hypersonic* jets, *boron hydrides* or "boranes" were a popular area of research because they generate more heat upon burning than an equal volume of hydrocarbon. This is significant for two reasons: (1) More heat means more energy, and (2) if a fuel is lighter, more can be carried. Both effects would result in an extended range of flight. Unfortunately, borane is thermally unstable, meaning that spontaneous explosion is difficult to control. Additionally, the fumes are toxic. While the use of boranes in *afterburners* was somewhat successful, the research turned out to be too expensive for its return on the dollar, and U.S. government funding in this area declined sharply in the late 1950s. (Borane studies did, however, lead to the discovery of novel mixed hydrides of carbon

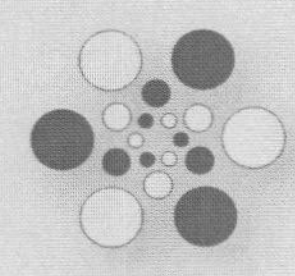

BORON AND PLANT NUTRITION

Boron is essential for the health of plants. This mineral is particularly important in regulating the structure of cell membranes. Root tip, leaf, and bud development depend on a constant supply of boron, which is easily washed out of the soil by excess watering. Drought conditions, however, also lead to boron deficiency. For most plants, an insufficiency can lead to anatomical changes and a failure to produce vigorous growth. An excellent source of boron for crops is organic matter like composted vegetable waste, which also contains elemental nutrients like nitrogen and potassium, whose functions in plant physiology are interrelated with those of boron. Regulation of boron fertilization is important, since excess can also be harmful to the environment, especially to beans and sweet potato crops.

More research into boron uptake and transport in plant systems will be important in the future: It is expected that an in-depth understanding of boron uptake in plants may lead to better knowledge of transport systems in all organisms, even the human kidney.

and boron, called carboranes. Carborane research has since led to the development of new reagents, catalysts, ceramics, polymers, and anticancer drugs.)

Recently, investigations into hydrogen as a clean fuel have led to ideas about boron augmentation of hydrogen fuels, but it does not seem to be feasible in practice—again because of the unstable nature of boranes. Another idea for powering vehicles involves combusting boron wire in the presence of pure oxygen. The B_2O_3 (boria) waste product could then be reprocessed into boron wire to be used as fuel in the same manner again. This research is still in its infancy, however, and may turn out to be economically unfeasible.

TECHNOLOGY AND CURRENT USES

Boron is a versatile element that possesses a number of important applications. Boric acid and borax have been used for centuries as antiseptics, based on possible antibacterial and antifungal properties. In insect extermination, boric acid is used to kill fleas, roaches, and house borers.

Boron and some of its compounds are used in rocket fuels. In fireworks, boron compounds give a brilliant green color. Compounds of boron are also used as water softeners.

Composite materials made of boron and aluminum are used in lightweight aerospace structures. Boron also contributes to the fiber-optics industry for use in a boron-germanium-doped silica material. Novel mixed hydrides of carbon and boron, called carboranes, have led to the development of new reagents, catalysts, ceramics, polymers, and anticancer drugs. Boron carbide is one of the substances used in nuclear control rods.

The isotope B-10 has a large slow *neutron-capture cross section* as it combines with neutrons, as shown in the following nuclear reaction:

$$^{10}_{5}B + ^{1}_{0}n \rightarrow ^{7}_{3}Li + ^{4}_{2}He,$$

where $^{4}_{2}He$ (a helium nucleus) is an *alpha particle*. A large neutron-capture cross section means that a substance can be used in neutron counters, neutron shields for nuclear workers, and as a *moderator* in nuclear reactors to control the rate of nuclear fusion. More than 80 percent of

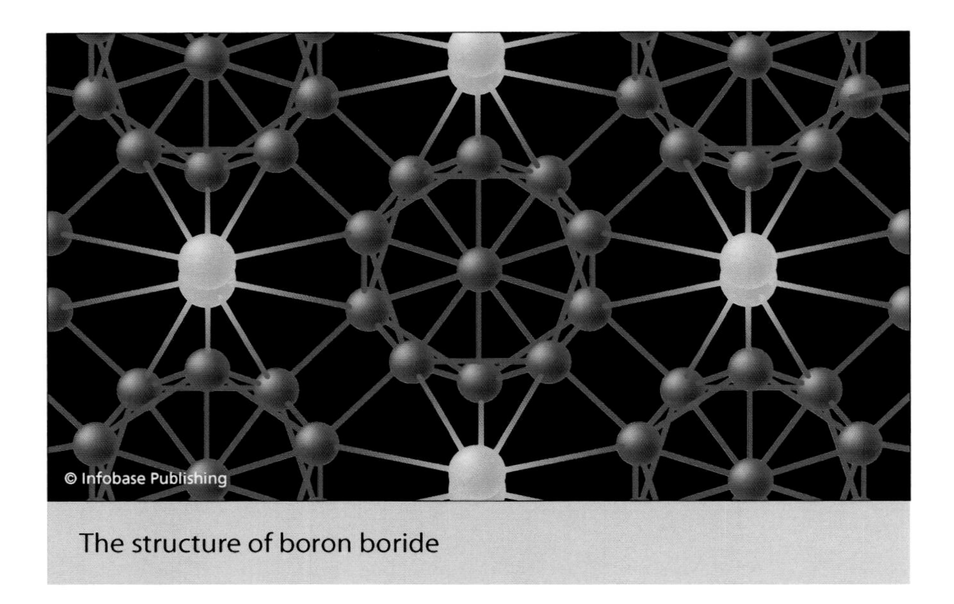

The structure of boron boride

natural boron, however, is B-11, which has a thermal neutron *cross section* that is about 80,000 times smaller than the cross section for B-10, so it is not suitable as a moderator. Consequently, samples of boron have to be enriched in B-10. Pure boron lacks the ductility and tensile strength to make rods, so boron carbide (B_4C) in molten aluminum is used.

Neutron counters use argon and BF_3 that is enriched with B-10. The alpha particle produced by the combination of B-10 with a neutron ionizes the argon gas, triggering the Geiger counter. The same reaction is used to treat brain tumors using the recoiling lithium and helium ions. In a similar manner, the neutron-absorbing power of an unknown sample can be used to estimate the boron content of the sample. In addition, both boron isotopes can be detected using *nuclear magnetic resonance* (NMR) techniques.

A new form of boron, "boron boride," produced at very high pressure in the laboratory, was discovered in 2009. This novel single-element compound consists of a lattice of clusters of 12 boron atoms joined with pairs of boron atoms, which gives it an ionic character such that charge can be transferred from the pairs to the lattice. Florida International University's Professor Jiuhua Chen, who initiated the project, says, "This has brought us a new understanding of the elements."

7

Silicon and Germanium

Silicon is element 14 and is the second most abundant element on Earth (after oxygen). It is a dark gray, lustrous solid with a density of 2.33 g/cm^3. Silicon is obtained mainly from quartz minerals and sand (which contains quartz). Germanium—element 32—is 52nd in abundance. Germanium is a grayish-white, rather hard and brittle, metallic-looking substance. It has a density of 5.32 g/cm^3. Germanium is principally found as a sulfide ore in association with sulfides of silver, lead, tin, antimony, and zinc.

The modern age of electronics began with the invention of the transistor in 1948. Silicon and germanium are the major components of semiconductors. The fact that they serve this role is due to their strategic positions in the periodic table. Just as carbon's position in the table gives carbon the properties that make it the basis of all living organisms, the

THE BASICS OF SILICON

Symbol: Si
Atomic number: 14
Atomic mass: 28.084; 28.086]
Electronic configuration: $[Ne]3s^23p^2$

T_{melt} = 2,577°F (1,414°C)
T_{boil} = 5,909°F (3,265°C)

Abundance
In Earth's crust 270,000 ppm

Isotope	Z	N	Relative Abundance
$^{28}_{14}Si$	14	14	92.23%
$^{29}_{14}Si$	14	15	4.68%
$^{30}_{14}Si$	14	16	3.09%

THE BASICS OF GERMANIUM

Symbol: Ge
Atomic number: 32
Atomic mass: 72.63
Electronic configuration:
$[Ar]4s^23d^{10}4p^2$

T_{melt} = 1,721°F (938°C)
T_{boil} = 5,131°F (2,833°C)

Abundance
In Earth's crust 1.4 ppm

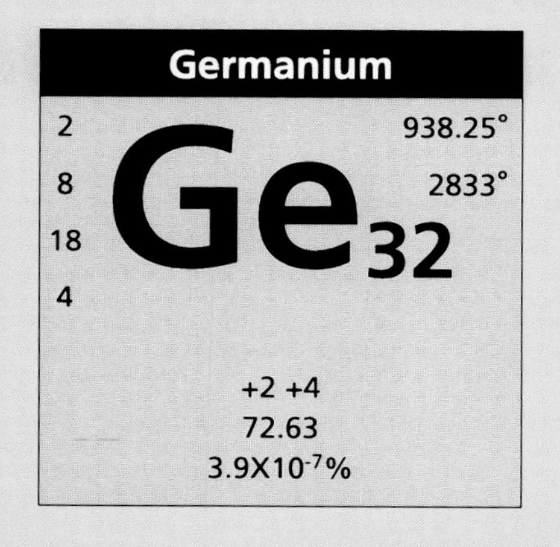

Isotope	Z	N	Relative Abundance
$^{70}_{32}Ge$	32	38	20.84%
$^{72}_{32}Ge$	32	40	27.54%
$^{73}_{32}Ge$	32	41	7.73%
$^{74}_{32}Ge$	32	42	36.28%
$^{76}_{32}Ge$	32	44	7.61%

positions of silicon and germanium in the same column—but as metalloids rather than nonmetals—give them the properties that make them semiconductors of electricity. An interesting contrast with the electrical conductivities of metals is that caused by an increase in temperature. With metals, electrical conductivity decreases at higher temperatures. In contrast, with silicon and germanium, electrical conductivity improves with increasing temperature. As shown in the diagram, there is a considerable gap in energy (the so-called *band gap*) in nonmetals between the valence band occupied by electrons and the conduction band in which electrons can flow. In metals, there is almost no gap at all. The gap in metalloids is intermediate between the two. At just slightly elevated temperatures, silicon's or germanium's valence electrons can move into the conduction band, and silicon or germanium can then begin to conduct electricity. Doping silicon or germanium with substances like gallium arsenide also increases their conductivity.

THE ASTROPHYSICS OF SILICON AND GERMANIUM

Most astrophysical silicon is synthesized in stars during oxygen burning. This process can occur in any very massive star (at least 15 times as massive as the Sun) that has evolved to the point that its core is mainly oxygen nuclei. The high mass is necessary to produce the high gravitational pressure and extreme temperature needed at the stellar core for fusion of such a heavy element as oxygen. About 10 percent of all stars meet this criterion. In such stars, silicon 28 and silicon 30 result from the fusion of two oxygen atoms when the star's core temperature reaches about 2×10^9 K.

$$^{16}_{8}O + ^{16}_{8}O \rightarrow ^{28}_{14}Si + ^{4}_{2}\alpha$$

$$^{16}_{8}O + ^{16}_{8}O \rightarrow ^{30}_{14}Si + 2^{1}_{1}p$$

While the alpha particles produced in the first process can also combine with existing ^{28}Si to make ^{32}Si, it is not until further core contraction causes the temperature to rise to around 4 billion K that the density leads to a high probability of alpha interaction.

$$^{28}_{14}Si + ^{4}_{2}\alpha \rightarrow ^{32}_{14}Si + \gamma$$

In addition to alpha particles, the ^{28}Si nuclei can capture protons and neutrons to make other nuclei up to and including the iron group elements in an overall process called "silicon burning."

The photons (γ) thus produced lead to *photodisintegration* of some of the silicon 28 nuclei, resulting in ^{24}Mg that can capture alphas to reverse the process and create more photons.

$$\gamma + {}^{28}_{14}\mathrm{Si} \rightarrow {}^{24}_{12}\mathrm{Mg} + \alpha \text{ or p's}$$

$$ {}^{24}_{12}\mathrm{Mg} + {}^{4}_{2}\alpha \rightarrow {}^{28}_{14}\mathrm{Si} + \gamma$$

When silicon burning is complete, the star ends its life in a supernova explosion that distributes its elements into space. Germanium, being heavier than iron, is synthesized in supernovae via the rapid capture by iron nuclei of a succession of neutrons, which is called the "r-process." An interesting puzzle has been brought to light in the study of isotopic ratios in silicon carbide stardust. It is believed that the Si-29 and Si-30 isotopes are most likely to be synthesized by neutron capture onto lighter nuclei. If that is the case, then the $^{29,30}Si$ abundances in dust from older stars should be smaller than the solar abundance, but the reverse seems to be true. The dilemma serves as a point of departure for further research.

DISCOVERY AND NAMING OF SILICON AND GERMANIUM

Silicon is found virtually everywhere. Quartz is composed of silica (silicon dioxide, SiO_2). Sand is mostly crushed quartz. Glass is made from sand. Although the manufacture of glass dates to at least 1500 B.C.E., as late as 1800 C.E., chemists were uncertain whether silica was a compound or an element. The French chemist Antoine-Laurent de Lavoisier (1743–94) believed that silica was the oxide of an as-yet undiscovered element. Gay-Lussac, Thénard, and Davy—the discoverers of boron—tried unsuccessfully to apply the same technique to silica that had worked to isolate boron from boric acid.

It remained for the Swedish chemist Jöns Jacob Berzelius (1779–1848) to isolate silicon. In 1824, by heating potassium in gaseous silicon tetrafluoride, Berzelius obtained a brown mass of impure silicon.

After repeated washings, he obtained an amorphous form of silicon—the name *silicon* being derived from the mineral *silica*. It was not until 1854, however, that crystalline silicon was obtained. This achievement is credited to Henri-Étienne Sainte-Claire Deville (1818–81), who was born in the West Indies and died near Paris. Deville is best remembered for his work on the production of aluminum. It was while working on techniques for obtaining larger quantities of aluminum that he succeeded in obtaining samples of crystalline silicon from silica impurities in the aluminum compounds with which he was working.

Silicon has the silvery *luster* that we associate with metals. Sainte-Claire Deville recognized, however, that silicon is not a metal, but a metalloid having properties intermediate between those of metals and nonmetals.

Dmitri Mendeleev predicted the existence of an element (which he called *eka-silicon*), unknown at that time, that would occupy the empty place in the periodic table subsequently occupied by germanium. In fulfillment of Mendeleev's prediction, germanium was discovered by the German chemist Clemens Alexander Winkler (1838–1904). Professor of Chemical Technology and Analytical Chemistry at the School of Mines in Freiberg, in 1885, Winkler and his students were analyzing the mineral argyrodite. Almost the entire mass of their sample was accounted for by known elements—but not a full 100 percent. Careful analysis showed that the small amount of remaining mass was a new element that they named *germanium* in honor of their native land, Germany.

It is interesting to note that, altogether, Mendeleev predicted the existence of at least four new elements. When those elements subsequently were discovered, all four were named in honor of the homelands of their discoverers: germanium for Germany, scandium for Scandinavia, gallium for France (Gallia), and polonium for Poland.

THE CHEMISTRY OF SILICON AND GERMANIUM

As the second most abundant element on Earth, silicon—in the form of *silicates*—is found in a number of common minerals. Silicates have important applications, including that of being components of ceramic materials. Thus, silicon in combination with oxygen is the most important way of finding and using silicon compounds.

As mentioned previously, quartz contains silicon dioxide, making silicon dioxide extremely abundant. It is interesting to contrast SiO_2 with its carbon analog, CO_2. Carbon dioxide consists of small molecules and is gaseous under normal conditions. Carbon forms a maximum of four chemical bonds to other atoms. No polymers or other chains are possible, because a CO_2 molecule contains two double bonds (in the form O=C=O), so that the carbon atom can form no additional bonds. In contrast, the Si–O bond in SiO_2 is a single bond, with the result that, in a quartz crystal, each silicon atom is bonded covalently to four oxygen atoms, and each oxygen atom is bonded to two silicon atoms. Consequently, silicon dioxide consists of a solid network of chains of alternating silicon and oxygen atoms.

Covalent silicon compounds take on a variety of forms that include the following examples: silicon tetrafluoride (SiF_4), a gas at normal temperatures; silicon tetrachloride ($SiCl_4$), a *volatile* liquid; and a number of silane hydrides, including SiH_4, Si_2H_6, Si_3H_8, and Si_4H_{10}, in analogy to the corresponding hydrocarbon compounds.

Silicon tetrafluoride reacts with hydrofluoric acid (HF) to yield fluorosilicic acid (H_2SiF_6), as shown in the following equation:

$$SiF_4 + 2\ HF \rightarrow H_2SiF_6\ (aq).$$

Fluorosilicic acid is a potent pesticide used to kill insects and rodents.

The first silane hydride, SiH_4, was synthesized in 1851 by Friedrich Wöhler (1800–82). Chemists were just beginning to understand organic chemistry, and Wöhler was intrigued by the possibility that silicon might exhibit a completely analogous chemistry. What Wöhler wanted to know was the answer to the following question: Could alcohols, aldehydes, ethers, carbohydrates, proteins, and other organic compounds be made in which the only difference would be the substitution of silicon atoms for carbon atoms? Wöhler's conclusion (and the conclusion of subsequent investigators along the same line) is that the answer is negative. The carbon-carbon bond is simply much stronger than the corresponding silicon-silicon bond, allowing even simple hydrocarbon chains containing 100 carbon atoms or more. Attempts to synthesize comparable silane chains showed that the chains tend to fall apart rather easily. The vitally important result is that life is based on carbon, not on silicon.

Combining silicon and carbon gives silicon carbide (SiC), a very hard solid similar to diamond. Silicon carbide is manufactured in large quantities, crushed into tiny particles, and used as an abrasive material—an example being the surface of a grinding wheel.

In organosilicon compounds, organic groups of atoms are bonded to silicon with Si–C bonding. First synthesized in the 1860s, several thousand such compounds are now known, and are the basis of silicone *polymers.* Usually, some of the oxygen atoms are replaced with other hydrocarbon groups such as a *methyl* group ($-CH_3$) or a *phenyl* group ($-C_6H_5$), leading to a greater variety of polymers with desired properties that can be designed beforehand. Sometimes the oxygen atoms provide cross-links between chains, resulting in a more rigid material. Silicone oil, for example, would not have any cross-links. Silicone rubber, on the other hand, would have at least a few cross-links.

The "+4" state is the stable oxidation state of germanium in compounds. Germanium compounds mimic their silicon analogs, and include germanium hydride (GeH_4), germanium chloride ($GeCl_4$), and germanium dioxide (GeO_2). These compounds tend not to have uses of their own, but are forms in which traces of germanium found in other ores can be separated from the parent ore. From $GeCl_4$, for example,

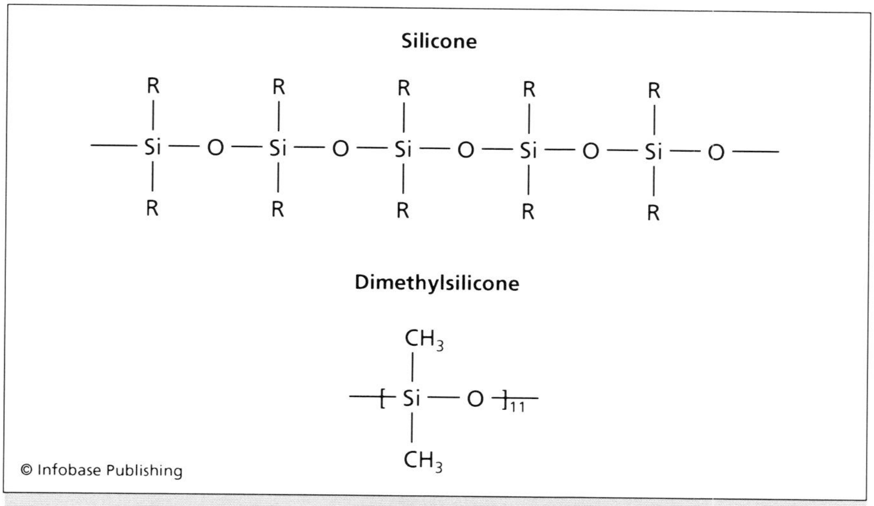

Silicone

Dimethylsilicone

Structure of a silicon polymer

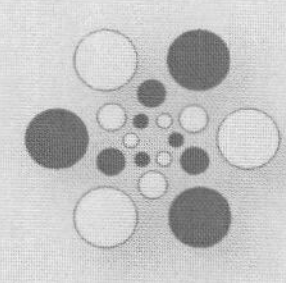

SAND AND GLASS

The earliest known human-made glass objects, found in Egypt and Mesopotamia, were beads that date from about 3500 B.C.E. Glass can be made by simply heating silicon dioxide (SiO_2)—commonly called silica—to about 2,000°C (3,630°F). Most sand contains iron, giving the glass a greenish tinge, which can be counteracted by adding arsenic. Pure white sand, which contains no iron, is in highest demand for glassmaking. Other additives are often included to give desired properties to the glass. Limestone, for example, can make glass less brittle, and lead can lend brilliance by changing the index of refraction.

Silica is found in sand and quartz all over the world and has been studied extensively. Its simple macroscopic appearance is deceptive. There are actually 22 different crystalline forms of SiO_2. Because some of these different forms are sold in rock shops, their names may be familiar to readers: rose quartz, amethyst, citrine, smoky quartz, jasper, opal, and obsidian. In petrified wood, silica crystals replace the organic components of the original woody material. Ultrapure synthetic quartz crystals are also produced in laboratories.

The main component of glass is silicon; other additives are included to give the glass color. *(Martine Oger/Shutterstock)*

pure germanium itself can be obtained by electrochemical reduction reactions.

SILICON-GERMANIUM SEMICONDUCTORS

Perfectly conducting materials allow electrons to move freely among the atoms of the material. At normal temperatures, there are no perfect conductors, but metals like silver and copper are very good.

Semiconductors, on the other hand, rely on the jumping of electrons between accessible energy levels for current to flow in the material. In a solid, these levels are more numerous than in a gas of the same species and form broad bands of available energies, allowing for large

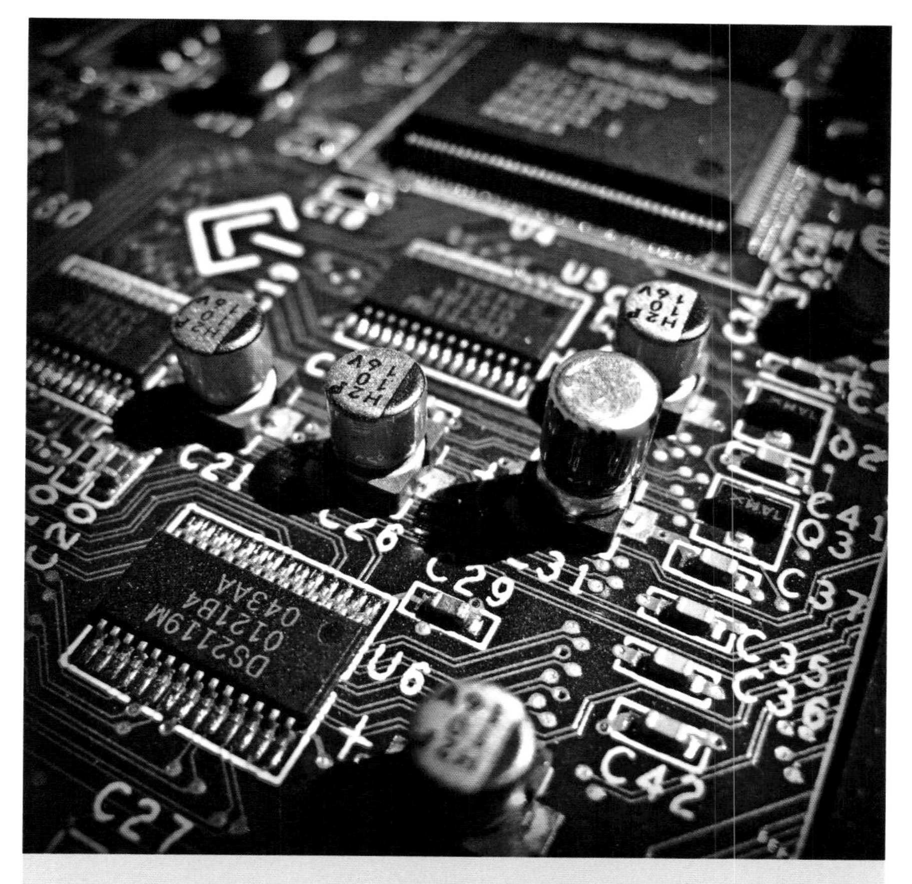

Most semiconductors used in electronic circuitry rely on silicon and germanium. *(Alessio Ponti/Shutterstock)*

congregations of electrons. These electrons undergo random thermal motion, but cannot jump to the next energy band if the gap between bands is much greater than the average thermal energy of the electrons. So current flow in semiconductors is strongly dependent on the ambient temperature. Electrons can also be made to jump between bands if they absorb energy in any other form, such as electromagnetic energy. An applied electric field or light absorption can make current flow in semiconductors. It is this property that makes them useful in circuits—small amounts of current can be made to flow at prescribed moments. This is an important property in transistors, optoelectronics, collision-avoidance radar, and solar cells.

The energy bandwidths and gaps are also dependent on the exact nature of the material. Silicon and germanium, for example, are good semiconductors, while carbon is not, though all three have similar electronic structure. (Note that they inhabit the same column in the periodic table.) The difference is that, in solid form, the gaps between the two highest energy levels in silicon and germanium are much smaller than in carbon, so electrons are able to transition between them with ease.

Though silicon has dominated the electronics industry for a decade or so, the latest hope for high-speed, inexpensive, low-noise microchips is a combination of silicon and germanium. Because of their similar chemical properties, germanium can be fairly easily introduced into a

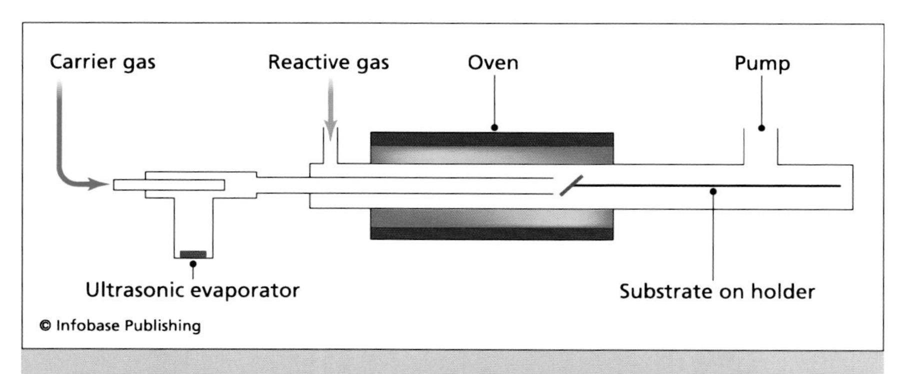

Chemical vapor deposition is a method for making thin films, as for semiconductor wafers.

silicon lattice by the technique of *chemical vapor deposition*. Germanium atoms are slightly larger than silicon atoms, which strains the structure somewhat, and lessens the band gap.

Silicon devices can deliver signals at a few billion cycles per second, but new silicon germanium (SiGe) technology has reached a speed of 500 billion cycles per second—though only at the extremely cold temperature of 4.5 K. SiGe circuits will be important in the ongoing development of space electronics, remote sensing, and data services.

TECHNOLOGY AND CURRENT USES

Ultrapure silicon is the main component of semiconductor parts. Other than workers in the semiconductor industry, few people probably have ever seen pure silicon or germanium. Both elements, however, have important applications in industry and the economy. For example, ferrosilicon (a combination of iron and silicon) is used in steelmaking. Silicon carbide is an important industrial abrasive. Quartz glass and ordinary glass (e.g., window glass) are made from silicon dioxide, which is also used to make Portland cement, comprising about 20 to 25 percent of that material. Silicates are also used as water softeners and cleaning agents. Mica is a silicate used in electrical insulating materials, paints, wallpaper, and Christmas tree "snow." Silicones are organic polymers based on silicon. They have rubberlike properties, and can be used for similar needs as rubber objects.

Asbestos is a silicate once widely used as a thermally insulating material (although less so currently because of its harmful effects on the lungs), and is still used in fireproof roofing shingles. Talc is a silicate used in soap, talcum powder, paint, and rubber. Feldspar is a silicate used in the glass and ceramic industries. Mixtures of silicates are used to manufacture ceramics, pottery, and bricks.

The principal use of germanium is in transistors. In addition, small amounts of germanium strengthen tin, aluminum, and magnesium alloys, and can enhance the speed of semiconductors.

It would be almost impossible to imagine a world without its modern electronics or without glass. There will always be a demand for silicon and germanium.

8

Arsenic and Antimony

Arsenic is element 33, with a density of 6.72 g/cm^3. It is fairly common (ranking 20th in abundance in Earth's crust). However, it is also very toxic. Arsenic has one stable isotope and 17 radioactive isotopes and exists in two allotropic forms—a yellow solid and a brittle, gray solid. The sulfide ores of most other elements—especially copper, iron, and lead—also contain arsenic sulfides. Minerals that are predominantly arsenic sulfides exist as well, usually as mixtures of realgar (As_4S_4) and orpiment (As_2S_3). Unfortunately, these arsenic-containing substances can leach arsenic into ground- and surface-water supplies. Consequently, the Environmental Protection Agency has issued a strict upper limit of 10 ppb on the amount of arsenic that can be present in municipal drinking-water systems.

Antimony is element 51. It is 63rd in abundance, with a density of 6.68 g/cm^3. Antimony has two stable isotopes and 31 radioactive

THE BASICS OF ARSENIC

Symbol: As
Atomic number: 33
Atomic mass: 74.92160
Electronic configuration:
[Ar]$4s^2 3d^{10} 4p^3$

T_{melt} = 1,503°F (817°C)
T_{boil} = *sublimes* at 1,137°F (614°C)

Abundance
In Earth's crust 2.1 ppm

Isotope	Z	N	Relative Abundance
$^{75}_{33}As$	33	42	100%

Arsenic

2	As 33	817t°
8		614s°
18		1400°
5		

+3 +5 -3
74.92160
$2.1X10^{-8}$%

THE BASICS OF ANTIMONY

Symbol: Sb
Atomic number: 51
Atomic mass: 121.760
Electronic configuration:
[Kr]$5s^2 4d^{10} 5p^3$

T_{melt} = 1,167°F (631°C)
T_{boil} = 2,889°F (1,587°C)

Abundance
In Earth's crust 0.2 ppm

Isotope	Z	N	Relative Abundance
$^{121}_{51}Sb$	51	70	57.21%
$^{123}_{51}Sb$	51	72	42.79%

Antimony

2	Sb 51	630.63°
8		1587°
18		
18		
5		

+3 +5 -3
121.760
$1.01X10^{-9}$%

isotopes. The most common antimony-containing mineral is stibnite (Sb_2S_3). In addition, antimony is often found in deposits of arsenic, copper, silver, and lead, and much antimony is recovered from lead ores such as galena (PbS). Worldwide, antimony is obtained from mines in China, Bolivia, and Idaho.

Before their gravitational collection into the solar system, all arsenic and antimony atoms were originally formed in galactic nucleosynthesis by either the astrophysical r-process or s-process. The r-process requires neutron capture at a relatively rapid rate, which generally occurs in supernovae explosions. Arsenic and antimony can also be formed when the nuclei of iron and heavier elements within stars capture slow-moving free neutrons—the s-process.

Both elements exist in "-3," "+3," and "+5" oxidation states, although the "+3" state is the most common ion in their salts. Arsenic in both the "+3" and "+5" states is used in *insecticides* and other kinds of *pesticides*.

DISCOVERY AND NAMING OF ARSENIC AND ANTIMONY

The word *arsenic* comes from the Greek, and means "yellow orpiment," orpiment being a common mineral consisting of arsenic sulfide. The ancient Greeks and Romans utilized slave labor to work their

Albertus Magnus is credited with the discovery of arsenic in the 13th century. *(The Granger Collection)*

Crystals of stibnite, the most common antimony-containing mineral, display a metallic sheen. *(Mark Schneider/Getty Images)*

mines. Arsenic salts were contaminants of the ores and known for their toxicity. Both the mining operations themselves and the subsequent smelting operations resulted in a heavy death toll. Salts of arsenic and the other nearby Group VA elements—phosphorus, antimony, and bismuth—were known also to medieval alchemists. The medieval Chinese knew about arsenic salts and used them as *rodenticides* and insecticides in farming and rice cultivation. Around the world, accidental arsenic poisoning was fairly common. In addition, arsenic was often the poison of choice when people wished to eliminate political rivals.

Ancient people knew about antimony sulfide, which was used as a cosmetic substance to darken eyebrows and enhance the eyes. Neither arsenic nor antimony, however, was obtained in pure form by ancient people. That success remained for the alchemists to accomplish. There is no record, however, of the year in which antimony was first isolated or what person should be credited with that accomplishment. The name *antimony* means "not alone," referring to its having been known for many centuries only in combination with other elements. The symbol *Sb* is derived from antimony's principal mineral, stibnite.

THE CHEMISTRY OF ARSENIC AND ANTIMONY

Arsenic and antimony follow the trend in Group VB of having five valence electrons, and, therefore, existing in compounds with oxidation states of "±3" and "±5." Upon descending the column, the sizes of atoms increase. An increase in sizes of atoms means that the outermost electrons are held less tightly in the atoms, which is a property of metals. Consequently, arsenic and antimony exhibit properties that are more metallic in nature than nitrogen and phosphorus do, both of which are elements that are strictly nonmetallic in nature.

In the yellow mineral orpiment, arsenic is present as arsenic sulfide (As_2S_3). Arsenic sulfide does not react with either dilute or concentrated hydrochloric acid. However, it is soluble in nitric acid, in which the sulfide ion is oxidized to sulfur, as shown in the following equation:

$$3\ As_2S_3\ (s) + 10\ HNO_3\ (aq) + 4\ H_2O\ (l) \rightarrow$$
$$6\ H_3AsO_4\ (aq) + 10\ NO\ (g) + 9\ S\ (s).$$

When an unknown substance suspected to be arsenic sulfide is treated with nitric acid, the appearance of yellow flakes of elemental sulfur is taken as confirmation of arsenic's presence.

In the black mineral stibnite, antimony is present as antimony sulfide (Sb_2S_3). Antimony sulfide is insoluble in dilute hydrochloric acid, but dissolves in concentrated hydrochloric acid, as shown by the following equation:

$$Sb_2S_3\ (s) + 6\ HCl\ (aq) \rightarrow 2\ SbCl_3\ (aq) + 3\ H_2S\ (g).$$

The product, hydrogen sulfide gas, is toxic and has a pungent odor often described as the "rotten egg odor."

The fact that antimony sulfide does dissolve in concentrated HCl but arsenic sulfide does not can be used to separate the two substances in a *qualitative analysis* scheme. Adding HCl to an unknown sample that could possibly contain both As_2S_3 and Sb_2S_3 will dissolve Sb_2S_3, leaving As_2S_3 unaffected. The presence of the Sb^{3+} (or $SbCl_3$) in solution can be confirmed by placing a few pieces of mossy tin in the solution. If Sb^{3+} is present (usually as $SbCl_3$), it will be reduced by the tin to elemental antimony, which will coat the tin with a black deposit. This reaction is shown in the following equation:

$$2\ SbCl_3\ (aq) + 3\ Sn\ (s) \rightarrow 2\ Sb\ (s) + 3\ SnCl_2\ (aq).$$

In the "+5" oxidation state, arsenic exists in compounds like arsenic acid (H_3AsO_4) or in arsenates, such as lead arsenate ($Pb_3(AsO_4)_2$). Lead arsenate is an insecticide. *Arsenites* (arsenic in the form of AsO_3^{3-}) and *arsenates* (arsenic in the form of AsO_4^{3-}) can both be reduced to arsine (AsH_3), in which arsenic is in the "−3" oxidation state. All forms of arsenic are poisonous to humans. For example, a dose of only 0.1 g of white arsenic (As_4O_6) is lethal to humans. At one time, arsenic was a common component of insecticides. In recent years, however, it has been phased out because of its toxicity to humans.

When pathologists suspect a case of arsenic poisoning, a victim's stomach contents are treated in such a way as to convert possible arsenic compounds that might be present into arsine. Any arsine that forms will decompose into metallic arsenic and hydrogen gas, as shown by the following equation:

$$2\ AsH_3\ (g) \rightarrow 2\ As\ (s) + 3\ H_2\ (g).$$

The elemental arsenic will exhibit a metallic mirrorlike appearance, and is evidence of arsenic's presence in quantities as low as 0.0005 g.

At normal temperatures, antimony does not react with water or with air. However, when antimony is heated to very high temperatures, it will burn with a brilliant flame to form Sb_4O_6, which reacts with sodium hydroxide to yield *antimonites*, as shown by the following equation:

$$Sb_4O_6\ (s) + 12\ NaOH\ (aq) \rightarrow 4\ Na_3SbO_3\ (aq) + 6\ H_2O\ (l).$$

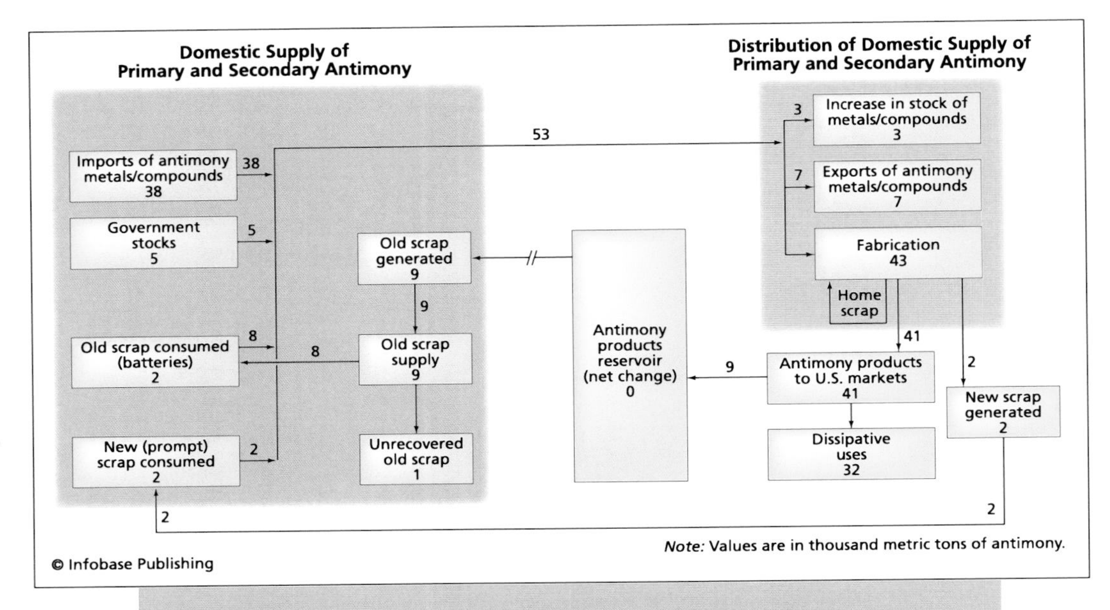

U.S. antimony materials flow from the year 2000 (values are in thousand metric tons)

Antimony reacts with all of the halogens to produce antimony trihalides. For example, chlorine gas and antimony react as shown by the following equation:

$$2\ Sb\ (s) + 3\ Cl_2\ (g) \rightarrow 2\ SbCl_3\ (s).$$

In addition, antimony trifluoride and antimony trichloride can be oxidized to pentahalides, in which antimony is in the "+5" oxidation state. These reactions are shown by the following equations:

$$SbF_3\ (s) + F_2\ (g) \rightarrow SbF_5\ (s),$$

$$SbCl_3\ (s) + F_2\ (g) \rightarrow SbCl_3F_2\ (s),$$

$$SbCl_3\ (s) + Cl_2\ (g) \rightarrow SbCl_5\ (s).$$

Antimony's principal use is to harden the metallic lead that is used in storage batteries. Because of its poisonous nature, recycling is important.

(continues on page 102)

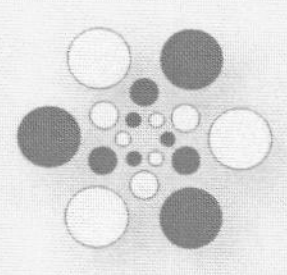

ARSENIC: A LEGENDARY POISON

Arsenic has been known as a toxin for centuries and is readily available: The mineral arsenic sulfide (As_2S_3), otherwise known as orpiment—a golden crystalline deposit—is commonly found around hot springs and volcanoes. It was used as an insect poison and to coat the tips of arrows in ancient China. As far back as the eighth century, an Arab chemist produced the inorganic compound arsenic trioxide, which is 500 times as toxic as pure arsenic. One could question why. That powder became the perfect tool for murder for more than 1,000 years. Its notoriety came to the forefront in Italy during the era of the Borgias, who were suspected of poisoning their enemies. Science had not yet the facility to detect the substance in the body, so accusations could seldom lead to conviction. Complicating the diagnosis was the coincidence that the symptoms of murder by arsenic were similar to those of cholera, which plagued Europe during the 19th century. The poison was made famous in *Arsenic and Old Lace,* the play and film in which two elderly sisters poison lonely old men, out of the kindness of their hearts, with their homemade elderberry wine.

(continues)

In Joseph Kesserling's play ***Arsenic and Old Lace,*** two elderly sisters murder lonely old men—out of the kindness of their hearts. *(George Karger/Pix Inc./Time Life Pictures/Getty Images)*

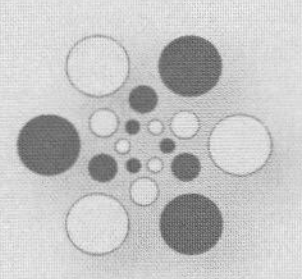

(continued)

Accidental ingestion of arsenic was more common in the past than today. Powdered orpiment made an excellent yellow pigment in paint and was particularly useful for a blend called "emerald green." This paint was employed by the Impressionists, and may have been the cause of illness and mental instability in some artists of that period. Green wallpaper and internal house paint containing arsenic were certainly responsible for sickening many people, notably American ambassador to Italy, Claire Booth Luce. During her stint there from 1953 to 1956, arsenic-laden paint flakes and powder that collected in her food and coffee caused her severe illness, though she did recover. Deadly arsenic vapors were also released from wallpaper through reactions with mold fungus. The inclusion of arsenic in artists' paint and wallpaper was mostly discontinued by 1900, but it has not been eliminated from all paints. It is currently used as a wood treatment and is a component in some laundry detergents, pesticides, and fertilizers. High amounts of arsenic have been found in some seafoods, beer, salt, tobacco, and bone meal. The seaweeds *kelp* and *hijiki* should not be

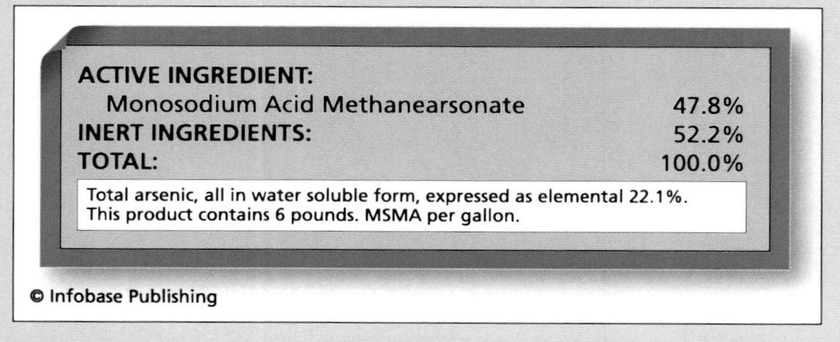

Pesticide products that contain arsenic in any form must include a substatement of the percentages of total arsenic and water-soluble arsenic calculated as elemental arsenic, as in the sample label shown.

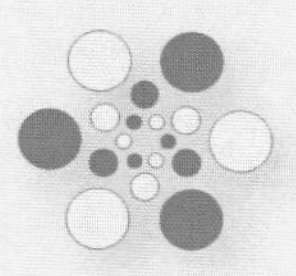

eaten on a regular basis because of their normally high arsenic content.

Owing to its abundance in Earth's crust, some arsenic does find its way into groundwater. Millions of shallow wells in Bangladesh that were dug in the 1970s to improve water quality have resulted in probably the largest mass poisoning in history. Arsenic-containing minerals washed down from the Himalayas that had lain undisturbed beneath the surface have now poisoned tens or even hundreds of thousands of the country's citizens. A solution to this devastating problem is not in sight.

Effects of arsenic poisoning, whether intentional or accidental, on the human system are diverse and can lead to death from

(continues)

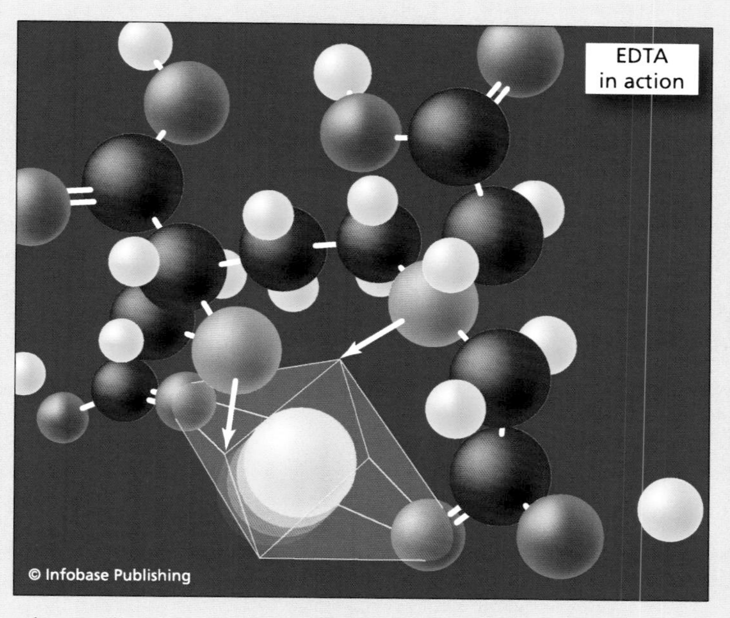

Chemically, the chelating agent is C-shaped. It surrounds a positively charged metal ion, making it inactive, and eventually removes it from the body.

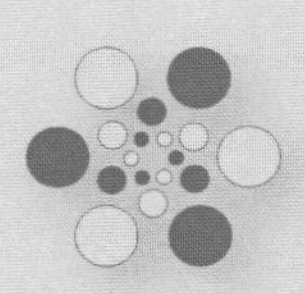

(continued)

multiple organ failures. Symptoms are extremely unpleasant and include stomach and intestinal pain and malfunction, convulsions, hoarseness and dry throat, headaches, and delirium. Continued exposure ultimately leads to death. The only known treatment is *chelation,* whereby a suitable chemical agent is introduced into the patient, usually orally. A chelating agent will bond easily with arsenic and subsequently be carried out in the urine. Studies have shown that garlic and other sulfur-containing foods have a mild chelating action for arsenic.

(continued from page 98)

FLAMEPROOFING WITH ANTIMONY

When one part antimony trioxide (Sb_2O_3) is combined with about four parts chlorine or bromine fire retardant, the flame resistance of the resulting compound is much higher than that of the halogen alone. This discovery led to the inclusion of antimony and halogens in mattresses, children's clothing and pajamas, toys, and aircraft and auto seat covers, as well as a wide variety of plastics, paints, and paper. It is important to note that the material is not thereby rendered incombustible. It can ignite, but the flame is immediately retarded. The process works on two levels. Upon combustion, antimony trihalides and oxyhalides form and trap free radicals in the vapor stage above the flame. Since free radicals propagate fire, sequestering them inhibits the burning. Additionally, Sb_2O_3 enhances the formation of a carbon-char surface layer, which doesn't burn easily and inhibits oxygen access to the material beneath it.

The addition of antimony to materials is a real fire-safety benefit, but there is another aspect that has gotten the attention of many child-protection groups: Antimony is poisonous. Because its structure is so similar to arsenic, it reacts with internal organs in much the same way, leading to symptoms of headaches, dizziness, vomiting, and even death when the dose is high or prolonged. Questions have arisen as to whether sufficient testing has been done regarding traversal of antimony out of the product at the surface, and how the human system might absorb it through inhalation and skin contact. Antimony is also a component in polyethylene terephthalate (PET) water bottles, and is known to leach into the water, especially at warm temperatures. Studies thus far seem to show that those particular levels are quite low, however—certainly below drinking-water regulatory guidelines.

TECHNOLOGY AND CURRENT USES

Because of arsenic's extensive use in the semiconductor industry, consumption of arsenic is much higher than that of antimony. The arsenide ion gallium, arsenic, and indium arsenide are used extensively in semiconductor components.

Several major uses of arsenic make use of its toxicity. White arsenic is used in the manufacture of insecticides and weed killers. Arsenic is the poison in some ant pastes and rodenticides. A compound of chromium, copper, and arsenic is used as a wood preservative. Some organic arsenic and antimony compounds are still the only known way to kill some harmful microorganisms (particularly certain parasites in Africa), but these treatments may be harmful to the patients.

Antimony's major use is as a flame retardant for plastics and fabrics. In addition, antimony compounds are components of glass, paints, and catalytic substances. Antimony is also used in lead storage batteries.

9

Tellurium and Polonium

Tellurium is element 52; it is a relatively soft, brown solid with a density of 6.24 g/cm^3. Tellurium's chemistry is very similar to the chemistry of sulfur and selenium, with the three of them comprising a *triad* of elements. Although tellurium is a fairly rare element (ranking number 72 in abundance), it has a large number of naturally occurring, stable isotopes and finds several applications in industry.

Polonium, on the other hand, has no stable isotopes. What little polonium does occur on Earth is formed by the radioactive decay of elements near it in the periodic table. Because polonium's isotopes all have short half-lives, only a negligible amount can ever accumulate in rocks containing uranium, thorium, and radium. The chemistry of polonium is similar to the chemistry of tellurium, although polonium is more metallic than tellurium is. Consequently, polonium tends to form salts as a metal does, for example, $Po(SO_4)_2$ and $Po(NO_3)_4$.

THE BASICS OF TELLURIUM

Symbol: Te
Atomic number: 52
Atomic mass: 127.60
Electronic configuration:
[Kr]$4d^{10}5s^{2}5p^{4}$

T_{melt} = 841°F (450°C)
T_{boil} = 1,810°F (988°C)

Abundance
In Earth's crust 0.001 ppm

Tellurium		
2	**Te** 52	449.51°
8		988°
18		
18		
6		
	+4 +6 -2	
	127.60	
	$1.57X10^{-8}$%	

Isotope	Z	N	Relative Abundance
$^{120}_{52}Te$	52	68	0.09%
$^{122}_{52}Te$	52	70	2.55%
$^{123}_{52}Te$	52	71	0.89%
$^{124}_{52}Te$	52	72	4.74%
$^{125}_{52}Te$	52	73	7.07%
$^{126}_{52}Te$	52	74	18.84%
$^{128}_{52}Te$	52	76	31.74%
$^{130}_{52}Te$	52	78	34.08%

THE BASICS OF POLONIUM

Symbol: Po
Atomic number: 84
Atomic mass: No stable isotopes
(Po-210 is the most common.)
Electronic configuration:
[Xe]$6s^{2}4f^{14}5d^{10}6p^{4}$

T_{melt} = 489°F (254°C)
T_{boil} = 1,764°F (962°C)

Polonium		
2	**Po** 84	254°
8		962°
18		
32		
18		
6	+2 +4	
	[209]	

Isotope	Z	N	Half-life
$^{206}_{84}Po$	84	122	8.8 days
$^{207}_{84}Po$	84	123	5.8 hours
$^{208}_{84}Po$	84	124	2.898 years
$^{209}_{84}Po$	84	125	102 years
$^{210}_{84}Po$	84	126	138 days

RARE ELEMENTS

Tellurium and polonium are both very rare elements on Earth and in space. Tellurium 130 is most likely an astrophysical r-process nuclide. The r-process requires neutron capture at a relatively rapid rate, which generally occurs in supernova explosions. On Earth, a few sulfide and selenide minerals also contain telluride. Tellurium is normally acquired as a mining by-product.

Polonium isotopes only form as daughters from the radioactive decay of unstable bismuth and radon isotopes as shown, for example, in the following:

$$^{210}_{83}\text{Bi} \rightarrow {}^{210}_{84}\text{Po} + {}^{0}_{-1}\text{e}$$

$$^{222}_{86}\text{Rn} \rightarrow {}^{218}_{84}\text{Po} + {}^{4}_{2}\text{He}.$$

Polonium 210 is the most important isotope. It undergoes alpha decay by the following reaction:

$$^{210}_{84}\text{Po} \rightarrow {}^{206}_{82}\text{Pb} + {}^{4}_{2}\text{He}.$$

In nuclear reactors, the alpha particles from this decay react with beryllium nuclei to produce neutrons which, in turn, initiate the nuclear *fission* of uranium, as illustrated by the following equations:

$$^{4}_{2}\text{He} + {}^{9}_{4}\text{Be} \rightarrow {}^{1}_{0}\text{n} + {}^{12}_{6}\text{C}$$

$$^{235}_{92}\text{U} + {}^{1}_{0}\text{n} \rightarrow {}^{89}_{36}\text{Kr} + {}^{144}_{56}\text{Ba} + 3{}^{1}_{0}\text{n}.$$

DISCOVERY AND NAMING OF TELLURIUM AND POLONIUM

The 18th century witnessed the discovery of many new elements, among them hydrogen, nitrogen, oxygen, chromium, manganese, cobalt, nickel, zinc, tungsten, and molybdenum. In the chronicles of the discoveries of elements, the 18th century closed with the discovery of tellurium.

Ores of unknown composition had been found in a gold mine in Romania. A metal obtained from the ores was at first thought perhaps to be antimony. In 1782, a Transylvanian mining official, Baron Franz Joseph Müller von Reichenstein (1740–1825), disputed this claim; he thought the substance was most likely bismuth sulfide. Müller, however, changed his mind the following year and decided that the unknown substance must be an unknown metal. He spent three years testing the substance and establishing its chemical and physical properties. Müller sent a sample of the unknown substance to the Swedish chemist Torbern Olaf Bergman (1735–84). Bergman agreed with Müller's conclusions but died of tuberculosis before he could do more confirmatory analyses.

Müller waited 12 years before continuing his study of the unknown substance. This time he sent a sample to the German chemist—and cofounder of analytical chemistry—Martin Heinrich Klaproth (1743–1817). In 1798, Klaproth succeeded in confirming that the unknown substance was in fact a new metallic element and gave Müller credit for its discovery. Klaproth named the new element *tellurium* from the Latin word *tellus,* meaning *earth.*

During the intervening years between Müller's initial isolation of tellurium and Klaproth's confirmation of tellurium's identity, the Hungarian scientist Paul Kitaibel (1757–1817) independently discovered tellurium in 1789. The issue of which scientist had priority of discovery—Müller or Kitaibel—was debated for several years. Although the two men never settled their dispute during Kitaibel's lifetime, historians have given Müller credit for tellurium's discovery.

In 1819, deposits of tellurium were discovered in the United States in Huntington, Connecticut. In 1848, a telluride ore was discovered in the vicinity of Fredericksburg, Virginia, and in 1857, tellurium was discovered in Georgetown, California. Ironically, the town of Telluride, Colorado, was named after tellurium, but tellurium was never found there. The mines of Telluride yielded gold, silver, copper, iron, lead, and zinc, but no tellurium.

The story of the discovery of the element radium is told in some detail in *The Alkali Metals and Alkaline Earths* in this set. The discovery of polonium is part of that same story. Uranium is the heaviest naturally occurring element on Earth. Among the heaviest elements, thorium occurs in somewhat greater abundance than uranium. All isotopes of both elements are radioactive. Beginning with these isotopes, several series of radioactive decays occur that produce trace amounts of the elements located in the periodic table between protactinium (Pa) and lead (Pb). (These decay series are described in more detail in *The Lanthanides and Actinides* in this set.) Polonium is one of the elements formed by these decays. Therefore, trace amounts of polonium are found in ores containing either uranium or thorium (or both).

In the late 1890s, the Polish-French chemist Marie Curie (1867–1934) worked diligently to separate small amounts of previously

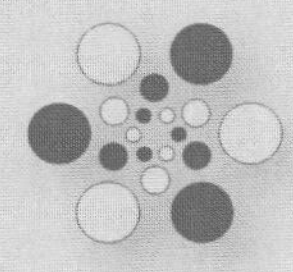

TELLURIUM IN SEMICONDUCTORS

Thin-film semiconductors, though more costly, are less bulky and much lighter in weight than the standard silicon semiconductors. Researchers have known since the 1980s how to make cadmium-tellurium (CdTe) films as thin as three microns for photovoltaic cells, which collect light in solar panels and turn it into electricity. Tellurium is an element of choice in solar cells because, when exposed to light, its conductivity increases. Layering tellurium with other elements that have different band gaps is ideal for solar power because sunlight has such a broad spread of wavelengths. The material for each layer is selected and doped to maximize light collection from as broad a spectrum as possible. Another semiconductor film proving efficient for solar energy collection combines zinc, manganese, and tellurium with oxygen impurities.

Relying on tellurium for solar panels may, however, restrict the extent to which solar energy can replace other more polluting types of energy production. Tellurium is extremely rare and is mainly acquired as a by-product of copper and gold mining. If all the tellurium produced per year globally were harnessed for solar power, it could provide about 2,400 megawatts of power. This is only about 0.5 percent of U.S. energy use per year.

Layering tellurium with bismuth and antimony also provides a material that allows for the conversion of heat (infrared wavelengths) to electricity and back with twice the efficiency of unknown elements from the uranium in pitchblende ore. One element she isolated in tiny quantities was radium. In 1898, a second element that she found in even tinier quantities was *polonium,* which she named for her native country Poland.

The discoveries of gallium and germanium have already been described. The existence of both of those elements had been predicted by Mendeleev who called them eka-aluminum and eka-silicon, since he expected them to resemble aluminum and silicon, respectively. Mendeleev

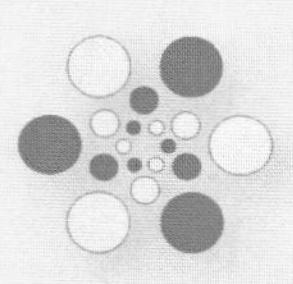

similar thermoelectric devices made of silicon. A distinct advantage of this technology is that the compound can be applied to specific tiny areas for pinpoint refrigeration. While expensive, it is already useful in spacecraft and high-tech laboratories. Scientists are looking into the idea of using such a material to recycle an automobile's waste heat to power the air conditioning system.

Tellurium is an element of choice in solar cells. *(Markus Gann/ Shutterstock)*

had also predicted the existence of an element that would be similar to tellurium, which he predicted would have an atomic weight of about 212. The isotopes of polonium, in fact, do have atomic weights close to 212, and the properties of polonium do resemble the properties of tellurium. The discovery of polonium, therefore, can be considered to be another success of Mendeleev's predictions. Of course, Mendeleev did not know about radioactivity in 1869, when he developed the periodic table, so radioactivity is a property of polonium that he could not have predicted.

THE LITVINENKO POISONING

Historically, polonium has not been regarded specifically as a poison in the sense that arsenic has. That changed on November 1, 2006, in London, England, when a former officer of the Russian Federal Security Service, Alexander Litvinenko, was fatally poisoned by ingestion of polonium 210 that had been slipped in his tea. Litvinenko was an outspoken critic of then Russian president, Vladimir Putin, and had recently accused him of responsibility for the murder of the journalist Anna Politkovskaya.

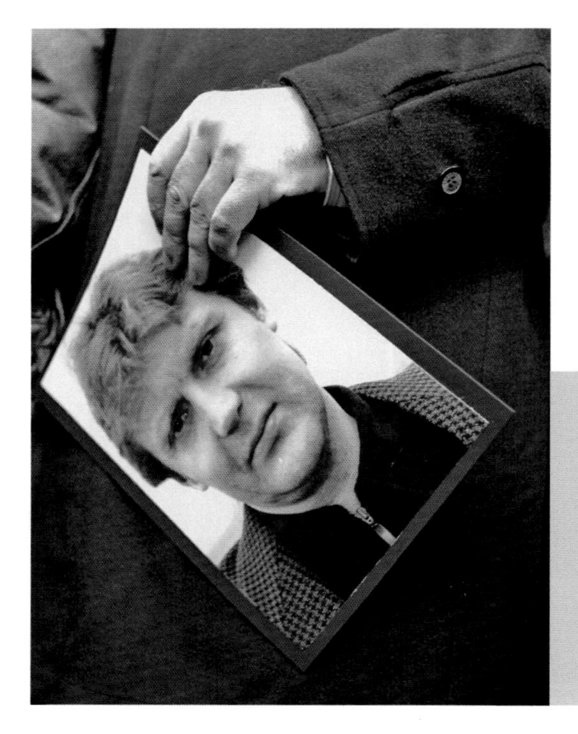

Alex Goldfarb, executive director of International Foundation for Civil Liberties, holds a picture of Alexander Litvinenko as he talks to the press on the first anniversary of Litvinenko's death by polonium poisoning. *(WILL WINTERCROSS/Bloomberg News/Landov)*

Polonium 210 was a clever choice of poison, as symptoms are delayed, and it is very difficult to detect. Polonium 210 decays by emitting alpha particles, but hospital radiation monitors typically detect only gamma rays. Original speculation was of thallium poisoning, as the symptoms are similar. (See chapter 3.) For acute doses of radiation, such as Litvinenko received, the alpha particles invade the cells and cause them to malfunction, leading to serious stomach and intestinal problems, fever, internal bleeding, hair loss, bacterial infections, and often death. Litvinenko died three weeks after his exposure. Before his death, he accused Putin of orchestrating his assassination.

It would be very difficult for anyone without access to a nuclear reactor to have acquired the amount of polonium used in the murder. In reactor science, polonium 210 is needed as a source of alpha particles, but its parent nuclide bismuth 210 is very scarce in nature, so it is manufactured by bombarding bismuth 209 with neutrons at the reactor site. Access to nuclear reactors is naturally severely restricted—Russian government agents would have been among the few allowed to enter or make use of polonium from a Russian reactor.

The prime suspects are Andrei Lugovoi and Dmitri Kovtun, two former KGB agents with whom Litvinenko had tea on that day. Traces of polonium were found in the teapot and cup and in several other locations in the hotel, as well as in the aircraft on which Lugovoi and Kovtun travelled. The two made it back to Russia, where they remain, despite a formal extradition request by the British government.

TECHNOLOGY AND CURRENT USES

Despite it being a fairly uncommon element, tellurium has several applications in industry. Tellurium is added to steel to increase ductility and to increase corrosion resistance. It is used in cracking of petroleum and as a brightener in electroplating baths. Tellurium compounds may also be used as coloring materials in glass.

Polonium serves as the high-activity alpha emitter that initiates the alpha-beryllium reaction to make neutrons that then stimulate uranium fission. It is also now notorious as an effective poison. In the future, polonium 210 may be useful to power long-distance space flight.

10

Conclusions and Future Directions

The periodic table of the elements is a marvelous tool, one that scientists have only begun to investigate. The key to its utility is its organization and the patterns it weaves. It can guide the eye and the mind to understand how far science has come and where human knowledge can lead.

SPECULATIONS ON FUTURE DEVELOPMENTS

It is important for scientists to think ahead, to attempt to guess what areas are ripe for investigation and which may be bound for oblivion. If one considers the recent remarkable leaps in information science, medicine, and particle physics that have materialized in the past century, it is clear that predictions are bound to be less impressive than eventualities, but there are some obvious starting points. Some of those, especially related to the metals and metalloids, are suggested here.

NEW PHYSICS

The near future of the metals and metalloids will most likely center on considerations regarding the environment and the promotion of cleaner energy. In particular, some environmental hazards need to be addressed in a timely manner.

While lead has been commonly acknowledged as unhealthy, legislation is not forthcoming. Ideas about federal regulations regarding lead pollution have been introduced in Congress, notably by President Barack Obama and Secretary of State Hillary Clinton, both of whom were U.S. senators at the time. Unfortunately, the problem is not widely understood. Efforts at education will be key to alleviating future health effects resulting from lead exposure.

The health risks of antimony also need to be investigated more rigorously. Questions have arisen as to whether sufficient testing has been done regarding traversal of antimony out of products that contain flame retardants, and how the human system might absorb antimony through inhalation and skin contact.

Energy concerns are ubiquitous in the media, and these should be addressed as speedily as possible. The promotion of aluminum recycling needs to be recognized as a crucial component in reducing global energy consumption, which could dramatically alleviate CO_2 emissions. Current aluminum recycling replaces only about 20 percent of annual aluminum consumption in the United States. Education is key for raising awareness about this issue.

Regarding new energy sources, the metals and metalloids will become more important than ever. Tellurium has become a crucial component in newer phase photovoltaic films, but relying on tellurium for solar panels may restrict the extent to which solar energy can replace other more polluting types of energy production. Tellurium is extremely rare and is mainly acquired as a by-product of copper and gold mining. If all the tellurium produced per year globally were harnessed for solar power, it could provide about 2,400 megawatts of power. This is only about 0.5 percent of U.S. energy use per year. For solar energy to produce a significant portion of that needed energy, different elements will need to take the place of tellurium.

Layering tellurium with bismuth and antimony provides a material that allows for the conversion of heat (infrared wavelengths) to electricity and back with twice the efficiency of similar thermoelectric devices made of silicon. A distinct advantage of this technology is that the compound can be applied to specific tiny areas for pinpoint refrigeration. While expensive, it is already useful in spacecraft and high-tech laboratories. Scientists are looking into the idea of using such a material to recycle an automobile's waste heat to power the air conditioning system.

New and improved nuclear reactors will also be on the rise in the near future. Molten lead-bismuth is the coolant of choice for the so-called Generation IV reactors, which are in development at the time of this writing. These would feature completely enclosed premanufactured modules with uranium alloy fuel cores surrounded by an outer shell containing the coolant. At the end of a typical useful lifetime of about 15 years, the module would be allowed to cool, and could then be shipped as a self-shielded, self-contained unit. Inherently safe, relatively compact, and economically competitive, such reactors will be extremely valuable for electricity production on small grids and in developing countries.

NEW CHEMISTRY

The chemistry of all of the post-transition metals and metalloids has been well studied. Probably the best prospects for new discoveries would be in the chemistry of boron and silicon. Boron chemistry has proven to be a rich field of research, as evidenced by the awarding of the 1979 Nobel Prize in chemistry to Herbert C. Brown and Georg Wittig for their development of boron-containing compounds that have become important reagents in organic synthesis. It is likely that new and useful boron compounds will continue to be made. Likewise, the field of synthetic polymers still has room for growth, so new silicone-based materials can be expected to appear on the market.

The discoveries of elements 113, 114, 115, and 116 have all been reported, but not confirmed independently by researchers at other laboratories. None of them has been named yet. Presumably, all of these elements will be post-transition metals, although it is difficult to foresee how chemists will ever produce sufficient quantities to study their physical and chemical properties in any detail. Nor would it be expected that sufficient quantities would be produced to find any commercial applications for them. For the time being, they represent extensions of the periodic table, but no more than that.

As for any new metalloids, element 117 is probably a metalloid, but its discovery has not yet been reported. Any isotopes of element 117 will be radioactive, with extremely short half-lives—probably on the order of nanoseconds. It, too, would represent an extension of the periodic table, but it is unlikely that scientists will ever be able to synthesize more than a few atoms at a time.

SI Units and Conversions

UNIT	QUANTITY	SYMBOL	CONVERSION
Base units			
meter	length	m	1 m = 3.2808 feet
kilogram	mass	kg	1 kg = 2.205 pounds
second	time	s	
ampere	electric current	A	
kelvin	thermodynamic temperature	K	1 K = 1°C = 1.8°F
candela	luminous intensity		
mole	amount of substance	mol	
Supplementary units			
radian	plane angle	rad	pi / 2 rad = 90°
steradian	solid angle	sr	
Derived units			
coulomb	quantity of electricity	C	
cubic meter	volume	m^3	1 m^3 = 1.308 $yards^3$
farad	capacitance	F	
henry	inductance	H	
hertz	frequency	Hz	
joule	energy	J	1 J = 0.2389 calories
kilogram per cubic meter	density	kg m^{-3}	1 kg m^{-3} = 0.0624 lb. ft^{-3}
lumen	luminous flux	lm	
lux	illuminance	lx	
meter per second	speed	m s^{-1}	1 m s^{-1} = 3.281 ft s^{-1}

UNIT	QUANTITY	SYMBOL	CONVERSION
meter per second squared	acceleration	$m\ s^{-2}$	
mole per cubic meter	concentration	$mol\ m^{-3}$	
newton	force	N	1 N = 7.218 lb. force
ohm	electric resistance	Ω	
pascal	pressure	Pa	$1\ Pa = \frac{0.145\ lb}{in^2}$
radian per second	angular velocity	$rad\ s^{-1}$	
radian per second squared	angular acceleration	$rad\ s^{-2}$	
square meter	area	m^2	$1\ m^2 = 1.196\ yards^2$
tesla	magnetic flux density	T	
volt	electromotive force	V	
watt	power	W	$1W = 3.412\ Btu\ h^{-1}$
weber	magnetic flux	Wb	

PREFIXES USED WITH SI UNITS		
PREFIX	SYMBOL	VALUE
atto	a	$\times 10^{-18}$
femto	f	$\times 10^{-15}$
pico	p	$\times 10^{-12}$
nano	n	$\times 10^{-9}$
micro	μ	$\times 10^{-6}$
milli	m	$\times 10^{-3}$
centi	c	$\times 10^{-2}$
deci	d	$\times 10^{-1}$
deca	da	$\times 10$
hecto	h	$\times 10^{2}$
kilo	k	$\times 10^{3}$
mega	M	$\times 10^{6}$
giga	G	$\times 10^{9}$
tera	T	$\times 10^{12}$

List of Acronyms

AGB	Asymptotic giant branch
CP	Chemically peculiar
DNA	Deoxyribonucleic acid
EDTA	Ethylenediaminetetraacetic acid
FDA	Food and Drug Association
ISM	Interstellar medium
LED	Light-emitting diode
RNA	Ribonucleic acid
SneI	Type-1 supernovae
TSP	Trisodium phosphate

Periodic Table of the Elements

Periodic Table of the Elements

Key: 18 (Atomic number) — Ar (Symbol) — 39.948 (Atomic weight)

Legend: Halogens; Metals; Nonmetals; Metalloids; Unknown

1 IA	2 IIA	3 IIIB	4 IVB	5 VB	6 VIB	7 VIIB	8 VIIIB	9 VIIIB	10 VIIIB	11 IB	12 IIB	13 IIIA	14 IVA	15 VA	16 VIA	17 VIIA	18 VIIIA
1 H [1.00784; 1.00811]																	2 He 4.0026
3 Li [6.938; 6.997]	4 Be 9.0122											5 B [10.806; 10.821]	6 C [12.0096; 12.0116]	7 N [14.00643; 14.00728]	8 O [15.99903; 15.99977]	9 F 18.9984	10 Ne 20.1797
11 Na 22.9898	12 Mg 24.3050											13 Al 26.9815	14 Si [28.084; 28.086]	15 P 30.9738	16 S [32.059; 32.076]	17 Cl [35.446; 35.457]	18 Ar 39.948
19 K 39.0983	20 Ca 40.078	21 Sc 44.9559	22 Ti 47.867	23 V 50.9415	24 Cr 51.9961	25 Mn 54.938	26 Fe 55.845	27 Co 58.9332	28 Ni 58.6934	29 Cu 63.546	30 Zn 65.38	31 Ga 69.723	32 Ge 72.63	33 As 74.9216	34 Se 78.96	35 Br 79.904	36 Kr 83.798
37 Rb 85.4678	38 Sr 87.62	39 Y 88.906	40 Zr 91.224	41 Nb 92.9064	42 Mo 95.96	43 Tc (98)	44 Ru 101.07	45 Rh 102.9055	46 Pd 106.42	47 Ag 107.8682	48 Cd 112.411	49 In 114.818	50 Sn 118.710	51 Sb 121.760	52 Te 127.60	53 I 126.9045	54 Xe 131.29
55 Cs 132.9055	56 Ba 137.327	57–71 lanthanides	72 Hf 178.49	73 Ta 180.948	74 W 183.84	75 Re 186.207	76 Os 190.23	77 Ir 192.217	78 Pt 195.08	79 Au 196.9666	80 Hg 200.59	81 Tl [204.382; 204.385]	82 Pb 207.2	83 Bi 208.9804	84 Po (209)	85 At (210)	86 Rn (222)
87 Fr (223)	88 Ra (226)	89–103 actinides	104 Rf (265)	105 Db (268)	106 Sg (271)	107 Bh (270)	108 Hs (277)	109 Mt (276)	110 Ds (281)	111 Rg (280)	112 Cn (285)	113 Uut* (284)	114 Uuq* (289)	115 Uup* (288)	116 Uuh* (293)	117 Uus* (294)	118 Uuo* (294)

Numbers in parentheses are atomic mass numbers of most stable isotopes.

Lanthanides	57 La 138.9055	58 Ce 140.116	59 Pr 140.908	60 Nd 144.24	61 Pm (145)	62 Sm 150.36	63 Eu 151.964	64 Gd 157.25	65 Tb 158.9254	66 Dy 162.500	67 Ho 164.9303	68 Er 167.26	69 Tm 168.9342	70 Yb 173.054	71 Lu 174.967
Actinides	89 Ac (227)	90 Th 232.0381	91 Pa 231.036	92 U 238.0289	93 Np (237)	94 Pu (244)	95 Am (243)	96 Cm (247)	97 Bk (247)	98 Cf (251)	99 Es (252)	100 Fm (257)	101 Md (258)	102 No (259)	103 Lr (262)

*Element is not officially named; designation is temporary.

Element Categories

Element Categories

Nonmetals

1	H	Hydrogen
6	C	Carbon
7	N	Nitrogen
8	O	Oxygen
15	P	Phosphorus
16	S	Sulfur
34	Se	Selenium

Halogens

9	F	Fluorine
17	Cl	Chlorine
35	Br	Bromine
53	I	Iodine
85	At	Astatine

Noble Gases

2	He	Helium
10	Ne	Neon
18	Ar	Argon
36	Kr	Krypton
54	Xe	Xenon
86	Ra	Radon

Metalloids

5	B	Boron
14	Si	Silicon
32	Ge	Germanium
33	As	Arsenic
51	Sb	Antimony
52	Te	Tellurium
84	Po	Polonium

Alkali Metals

3	Li	Lithium
11	Na	Sodium
19	K	Potassium
37	Rb	Rubidium
55	Cs	Cesium
87	Fr	Francium

Alkaline Earth Metals

4	Be	Beryllium
12	Mg	Magnesium
20	Ca	Calcium
38	Sr	Strontium
56	Ba	Barium
88	Ra	Radium

Post-Transition Metals

13	Al	Aluminum
31	Ga	Gallium
49	In	Indium
50	Sn	Tin
81	Tl	Thallium
82	Pb	Lead
83	Bi	Bismuth

Transactinides

104	Rf	Rutherfordium
105	Db	Dubnium
106	Sg	Seaborgium
107	Bh	Bohrium
108	Hs	Hassium
109	**Mt**	**Meitnerium**
110	**Ds**	**Darmstadtium**
111	**Rg**	**Roentgenium**
112	**Cn**	**Copernicium***
113	**Uut**	**Ununtrium***
114	**Uuq**	**Ununquadium***
115	**Uup**	**Ununpentium***
116	**Uuh**	**Ununhexium***
117	**Uus**	**Ununseptium***
118	**Uuo**	**Ununoctium***

Transition Metals

21	Sc	Scandium	39	Y	Yttrium	72	Hf	Hafnium
22	Ti	Titanium	40	Zr	Zirconium	73	Ta	Tantalum
23	V	Vanadium	41	Nb	Niobium	74	W	Tungsten
24	Cr	Chromium	42	Mo	Molybdenum	75	Re	Rhenium
25	Mn	Manganese	43	Tc	Technetium	76	Os	Osmium
26	Fe	Iron	44	Ru	Ruthenium	77	Ir	Iridium
27	Co	Cobalt	45	Rh	Rhodium	78	Pt	Platinum
28	Ni	Nickel	46	Pd	Palladium	79	Au	Gold
29	Cu	Copper	47	Ag	Silver	80	Hg	Mercury
30	Zn	Zinc	48	Cd	Cadmium			

Lanthanides

57	La	Lanthanum	62	Sm	Samarium	67	Ho	Holmium
58	Ce	Cerium	63	Eu	Europium	68	Er	Erbium
59	Pr	Praseodymium	64	Gd	Gadolinium	69	Tm	Thulium
60	Nd	Neodymium	65	Tb	Terbium	70	Yb	Ytterbium
61	Pm	Promethium	66	Dy	Dysprosium	71	Lu	Lutetium

Actinides

89	Ac	Actinium	94	Pu	Plutonium	99	Es	Einsteinium
90	Th	Thorium	95	Am	Americium	100	Fm	Fermium
91	Pa	Protactinium	96	Cm	Curium	101	Md	Mendelevium
92	U	Uranium	97	Bk	Berkelium	102	No	Nobelium
93	Np	Neptunium	98	Cf	Californium	103	Lr	Lawrencium

Note: The organization of eriodic table of the elements, while useful to chemists and physicists, may be confusing to nonscientists in that some groupings of similar elements appear as vertical columns (halogens, for example), some as horizontal rows (lanthanides, for example), and some as a combination of both (nonmetals).

The table of element categories is intended as a quick reference sheet to easily determine which elements belong to which groups.

* Element is not officially named; designation is temporary.

Chronology

ca. fourth century B.C.E. Bronze is used in ancient Mesopotamia.

Lead is used in ancient Egypt and China.

ca. second century B.C.E. Glass is being manufactured around the world in places like the Mediterranean, Middle East, and China.

1400s C.E. Bismuth is used by medieval alchemists.

1735 The Swedish chemist Torbern Olaf Bergman is born on March 20 in Katrineberg, Sweden.

1740 Transylvanian mining official Baron Franz Joseph Müller von Reichenstein is born on July 1 in Nagyszeben (Hermannstadt) Transylvania (now Hungary).

1743 German chemist Martin Heinrich Klaproth is born on December 1 in Wernigerode, Germany.

1757 The Hungarian scientist Paul Kitaibel is born on February 3 in Mattersburg, Austria.

1777 The French chemist Louis-Jacques Thénard is born on May 4 in La Louptière, Aube, France.

The Danish scientist Hans Christian Ørsted is born on August 14 in Rudkøbing, Denmark.

1778 The French chemist Louis-Joseph Gay-Lussac is born on December 6 in Saint-Leonard-de-Noblat, France.

The English chemist Humphrey Davy is born on December 17 in Penzance, Cornwall, England.

1779 The Swedish chemist Jöns Jacob Berzelius is born on August 20 in Väversunda, Östergötland, Sweden.

1780s Müller von Reichenstein reports tentative discovery of tellurium.

1782 Baron von Reichenstein begins studies that lead to the isolation of tellurium.

1784 Torbern Bergman dies on July 8 in Medevi, Sweden.

1789 Paul Kitaibel independently discovers tellurium.

1798 Martin Klaproth confirms Müller's discovery of tellurium.

1799 German physicist Ferdinand Reich is born on February 19 in Bernburg, Germany.

1800 German chemist Friedrich Wöhler is born on July 31 in Eschersheim, Frankfurt am Main, Germany.

1808 Gay-Lussac and Thénard in France and Davy in England independently isolate boron.

1817 Martin Klaproth dies on January 1 in Berlin, Germany.

Paul Kitaibel dies on December 13 in Budapest, Hungary.

1818 The French chemist Henri-Étienne Sainte-Claire Deville is born on March 9 on the island of St. Thomas in the West Indies.

1819 Tellurium is discovered in the state of Connecticut.

1820 Ørsted discovers the principle of electromagnetism during a college lecture.

French chemist Claude-Auguste Lamy is born on June 15 in Ney, France.

1824 Berzelius isolates silicon.

German chemist Hieronymus Richter is born on November 21 in Dresden, Germany.

1825 Ørsted isolates aluminum.

Baron von Reichenstein dies on October 12 in Vienna, Austria.

1827 Wöhler prepares pure aluminum.

1829 Sir Humphrey Davy dies on May 29 in Geneva, Switzerland.

1832 British scientist William Crookes is born on June 17 in London, England.

1834 Russian chemist Dmitri Mendeleev is born on January 27 in Tobolsk, Siberia.

1838 The French chemist Paul-Emile Lecoq de Boisbaudran is born on April 18 in Cognac, France.

The German chemist Clemens Alexander Winkler is born on December 26 in Freiberg, Germany.

1848 Jöns Jacob Berzelius dies on August 7 in Stockholm, Sweden.

1850 Louis-Joseph Gay-Lussac dies on May 9 in Saint-Leonard-de-Noblat, France.

1851 Hans Christian Ørsted dies on March 9 in Copenhagen, Denmark.

Friedrich Wöhler synthesizes the first silane hydride, SiH_4.

1854 Henri Deville prepares pure silicon and recognizes it as a metalloid.

1857 Louis-Jacques Thénard dies on June 21 in Paris, France.

1860s Borax is discovered in southern California.

1861 Sir William Crookes and Claude-Auguste Lamy independently discover thallium.

1863 The French chemist Paul-Louis-Toussaint Héroult is born on April 10 in Saint-Benin, France.

Reich and Richter isolate indium.

The American chemist Charles Martin Hall is born on December 6 in Thompson, Oregon.

1867 Polish-French chemist Marie Curie is born on November 7 in Warsaw, Poland.

1869 Mendeleev publishes his periodic table of the elements.

1875 Paul-Emile Lecoq de Boisbaudran isolates gallium.

1878 Claude-Auguste Lamy dies on March 20 in Paris, France.

1881 Henri Deville dies on July 1 in Boulogne-sur-Seine, France.

1882 Ferdinand Reich dies on April 27 in Freiberg, Germany.

Friedrich Wöhler dies on September 23 in Göttingen, Germany.

1885 Winkler isolates germanium.

1886 Charles Hall and Paul Héroult independently develop a process for producing aluminum metal on a commercial scale.

Clemens Winkler discovers germanium.

1897 German chemist Georg Wittig is born on June 16 in Berlin, Germany.

1898 Hieronymus Richter dies on September 25 in Freiberg, Germany.

1904 Clemens Winkler dies on October 8 in Dresden, Germany.

1907 Dmitri Mendeleev dies on January 20 in St. Petersburg, Russia.

1911 Marie Curie receives the Nobel Prize in Chemistry for her discovery of radium and polonium.

1912 American chemist Herbert C. Brown is born on May 12 in London, England, to Ukrainian parents.

Paul-Emile de Boisbaudran dies on May 28 in Paris, France.

1914 Paul Héroult dies on May 9 aboard a yacht in the Mediterranean Sea.

Charles Hall dies on December 27 in Daytona, Florida.

1919 Sir William Crookes dies on April 4 in London, England.

1924 American chemist M. Frederick Hawthorne is born on August 24 in Fort Scott, Kansas.

1934 Marie Curie dies on July 4 in Paris, France.

1942 Herbert Brown discovers sodium borohydride.

1948 Bell Laboratories unveils the transistor on June 30.

1979 Herbert Brown and Georg Wittig each receive one-half of the Nobel Prize in chemistry for their development of the use of boron- and phosphorus-containing compounds.

1981 Herbert Brown receives the Priestley Medal, the highest award of the American Chemical Society.

1987 Georg Wittig dies on August 26 in Heidelberg, Germany.

1998 The discovery of element 114 (as yet unnamed) is announced by workers at the Nuclear Institute at Dubna, Russia.

2000 The discovery of element 116 (as yet unnamed) is announced by workers at the Nuclear Institute at Dubna, Russia.

2003 The discoveries of elements 113 and 115 (both as yet unnamed) are announced by workers at the Nuclear Institute at Dubna, Russia.

2004 Herbert Brown dies on December 19 in Lafayette, Indiana.

2006 Researchers at the Lawrence Livermore National Laboratory in California and the Joint Institute for Nuclear Research in Dubna, Russia, report discovery of element 118, which is located just below radon in the periodic table.

2009 Fred Hawthorne receives the Priestley Medal from the American Chemical Society on March 24 in Salt Lake City, Utah.

Researchers at Florida International University report the discovery of single-element compound, boron boride (B_{28}).

2010 The discovery of element 117 (as yet unnamed) is announced by researchers at the Joint Institute for Nuclear Research in Dubna, Russia.

Glossary

acid a type of compound that contains hydrogen and dissociates in water to produce hydrogen ions.

acid chloride an organic chemical compound that has been formed by having a chlorine atom replace the –OH portion of a –COOH group.

actinide the elements ranging from thorium (atomic number 90) to lawrencium (number 103); they all have two outer electrons in the 7s subshell plus increasingly more electrons in the 5f subshell.

acute having a rapid and intense onset, as in an acute radiation dose.

affinity See electron affinity.

afterburner used in jet engines to provide additional thrust by burning fuel with some exhaust gases.

aldehyde an organic chemical compound that contains the –CHO group.

alkali metal the elements in column IA of the periodic table (exclusive of hydrogen); they all are characterized by a single valence electron in an s subshell.

alkaline earth metal the elements in column IIA of the periodic table; they all are characterized by two valence electrons that fill an s subshell.

alkene an organic compound in which at least two carbon atoms are linked together by a double bond.

alpha decay a mode of radioactive decay in which an alpha particle—a nucleus of helium 4—is emitted; the daughter isotope has an atomic number two units less than the atomic number of the parent isotope, and a mass number that is four units less.

amphoteric a chemical compound that will dissolve in either an acid or a base.

anion an atom with one or more extra electrons, giving it a net negative charge.

anode the site in an electrochemical cell where oxidation occurs; the anode is positive in an electrolytic cell and is negative in a galvanic cell.

antineutrino an elementary particle identical to the neutrino, but having opposite spin.

aqueous describing a solution in water.

asteroids rocky interplanetary objects, mostly confined to a region between Mars and Jupiter.

asymptotic giant branch (AGB) an area of the Hertzsprung-Russell diagram above the main sequence line, where some high-mass stars (AGB stars) are mapped for luminosity and temperature.

atom the smallest part of an element that retains the element's chemical properties; atoms consist of protons, neutrons, and electrons.

atomic mass the mass of a given isotope of an element—the combined masses of all its protons, neutrons, and electrons.

atomic number the number of protons in an atom of an element; the atomic number establishes the identity of an element.

atomic weight the mean weight of the atomic masses of all the atoms of an element found in a given sample, weighted by isotopic abundance.

band gap the separation between adjacent energy levels in a crystal.

base a substance that reacts with an acid to give water and a salt; a substance that, when dissolved in water, produces hydroxide ions.

beta decay a mode of radioactive decay in which a beta particle—an ordinary electron—is emitted; the daughter isotope has an atomic number one unit greater than the atomic number of the parent isotope, but the same mass number.

biological magnification the increase in concentration of a substance moving up the food chain.

block a major section of the periodic table defined by the kinds of outermost, or valence, electrons the atoms in that section possess.

The first two columns of the table are called the "s" block. The middle 10 columns are called the "d" block. The six far right-hand columns are called the "p" block. The two rows at the bottom of the table are called the "f" block.

borane See boron hydride.

boron hydride a compound that contains boron and hydrogen.

cathode the site in an electrochemical cell where reduction occurs; cathode is negative in an electrolytic cell and is positive in a galvanic cell.

cation an atom that has lost one or more electrons to acquire a net positive charge.

chelate a chemical configuration in which a metal ion is an integral and central component binding a large organic molecule.

chelation See chelate.

chemical change a change in which one or more chemical elements or compounds form new compounds; in a chemical change, the names of the compounds change.

chemical vapor deposition a process for making thin films by evaporation of the desired material onto a substrate.

chronometer any device that keeps track of time in a very precise manner.

complex ion any ion that contains more than one atom.

compound a pure chemical substance consisting of two or more elements in fixed, or definite, proportions.

covalent bond a chemical bond formed by sharing valence electrons between two atoms (in contrast to an ionic bond, in which one or more valence electrons are transferred from one atom to another atom).

critical point temperature the temperature of a pure substance at which the liquid and gaseous phases become indistinguishable.

cross section in nuclear and atomic reactions, the probability for the reaction to occur.

diatomic molecule a molecule that contains two atoms.

dimer a chemical compound formed by two identical molecules linking together.

dioxin a toxic chlorine-containing hydrocarbon found in some weed killers.

ductility the ability of certain metals to be able to be drawn into thin wires without breaking.

electrical conductivity the ability of a substance, such as a metal, or a solution to conduct an electrical current.

electrolysis the process of causing a chemical reaction to occur by passing an electrical current through a solution.

electrolyte a substance that, when dissolved in water, makes the solution electrically conducting.

electron a subatomic particle found in all neutral atoms; possesses the negative charges in atoms.

electron affinity the energy released when a neutral atom gains an extra electron to form a negative ion.

electronegativity the relative tendency of the atoms of an element to attract the electrons between two atoms in a chemical bond.

electronic configuration a description of the arrangement of the electrons in an atom or ion, showing the numbers of electrons occupying each subshell.

electrostatic the type of interaction that exists between electrically charged particles; electrostatic forces attract particles together if the particles have charges of opposite sign, while the forces cause particles that have charges of like sign to repel each other.

element a pure chemical substance that contains only one kind of atom.

fallout radioactive particles deposited from the atmosphere from either a nuclear explosion or a nuclear accident.

family See group.

fission See nuclear fission.

fluorosis a toxic condition resulting from excess fluoride acting on a biological system.

furan a chemical compound consisting of a five-member ring—four carbon atoms and one oxygen atom—with two carbon-carbon double bonds.

gamma decay a mode of radioactive decay in which a very high-energy photon of electromagnetic radiation—a gamma ray—is emitted; the daughter isotope has the same atomic number and mass number as the parent isotope, but lower energy.

gamma ray a high-energy photon.

greenhouse gas (GHG) any gas that absorbs and reemits infrared radiation (heat).

group the elements that are located in the same column of the periodic table; also called a family, elements in the same column have similar chemical and physical properties.

half-life the time required for half of the original nuclei in a sample to decay; during each half-life, half of the nuclei that were present at the beginning of that period will decay.

Hall-Héroult cell an electrolytic cell used in industry to produce aluminum metal from its ores.

halogen the elements in column VIIB of the periodic table; all of them share a common set of seven valence electrons in an nth energy level such that their outermost electronic configuration is ns^2np^5.

Hertzsprung-Russell (HR) diagram used in astrophysics, a graph that plots luminosity versus surface temperature of a star.

hydrocarbon an organic chemical compound or group of atoms that contains only carbon and hydrogen.

hypersonic referring to speeds equal to or exceeding Mach 5.

inert an element that has little or no tendency to form chemical bonds; the inert gases are also called *noble gases.*

ingot a piece of metal or other material that is cast into a shape for further processing.

insecticide a poisonous substance intended to kill unwanted insects.

ion an atom or group of atoms that have a net electrical charge.

ion exchange is a reversible chemical reaction in which an ion from solution is exchanged for a similarly charged ion attached to an immobile solid.

ionic bond a strong electrostatic attraction between a positive ion and a negative ion that holds the two ions together.

isotope a form of an element characterized by a specific mass number; the different isotopes of an element have the same number of protons but different numbers of neutrons, hence different mass numbers.

ketone an organic chemical compound in which a –CO group is attached to two hydrocarbon groups.

lanthanide the elements ranging from cerium (atomic number 58) to lutetium (atomic number 71); they all have two outer electrons in the 6s subshell plus increasingly more electrons in the 4f subshell.

luster the shininess associated with the surfaces of most metals.

magic number referring to nuclei that have closed energy shells. Groupings of 2, 8, 20, 50, 28, 82, and 126 make filled nuclear shells.

main group element an element in one of the first two columns or one of the right-hand six columns of the periodic table; distinguished from transition metals, which are located in the middle of the table, and from rare earths, which are located in the lower two rows shown apart from the rest of the table.

malleability the ability of a substance such as a metal to change shape without breaking; metals that are malleable can be hammered into thin sheets.

mass a measure of an object's resistance to acceleration; determined by the sum of the elementary particles comprising the object.

mass number the sum of the number of protons and neutrons in the nucleus of an atom. (See also isotope.)

metal any of the elements characterized by being good conductors of electricity and heat in the solid state; approximately 75 percent of the elements are metals.

metalloid (also called semimetal) any of the elements intermediate in properties between the metals and nonmetals; the elements in the periodic table located between metals and nonmetals.

muon an elementary particle identical to the electron, except that it has 207 times the mass.

nebula an interstellar collection of gas and dust that reflects or absorbs starlight to form luminous or dark clouds, respectively.

neutron the electrically neutral particle found in the nuclei of atoms.

noble gas any of the elements located in the last column of the periodic table—usually labeled column VIII or 18, or possibly column 0.

nonmetal the elements on the far right-hand side of the periodic table that are characterized by little or no electrical or thermal conductivity, a dull appearance, and brittleness.

nova a star that displays a sudden extreme increase in brightness, then gradually fades.

nuclear fission the process in which certain isotopes of relatively heavy atoms such as uranium or plutonium break apart into fragments of comparable size; accompanied by the release of large amounts of energy.

nuclear fusion the process in which certain isotopes of relatively light atoms such as hydrogen or helium can combine to form heavier isotopes; accompanied by the release of large amounts of energy.

nucleon a particle found in the nucleus of atoms; a proton or a neutron.

nucleosynthesis the process of building up atomic nuclei from protons and neutrons or from smaller nuclei.

nucleus the small, central core of an atom.

optoelectronics an area of research involving the use of light in electronic devices.

order of magnitude differing by a factor of 10.

organoborane an organic chemical compound that contains a boron atom.

orpiment a common mineral consisting of arsenic sulfide.

oxidation an increase in an atom's oxidation state; accomplished by a loss of electrons or an increase in the number of chemical bonds to atoms of other elements. (See also oxidation state.)

oxidation-reduction reaction a chemical reaction in which one element undergoes an increase in its oxidation state and another element undergoes a decrease in oxidation state.

oxidation state a description of the number of atoms of other elements to which an atom is bonded. A neutral atom or neutral group of atoms of a single element is defined to be in the zero oxidation state. Otherwise, in compounds, an atom is defined as being in a positive or negative oxidation state, depending upon whether the atom is bonded to elements that, respectively, are more or less electronegative than that atom is.

period any of the rows of the periodic table; rows are referred to as periods because of the periodic, or repetitive, trends in the properties of the elements.

periodic repeating at even intervals.

periodic table an arrangement of the chemical elements into rows and columns such that the elements are in order of increasing atomic number, and elements located in the same column have similar chemical and physical properties.

photodisintegration the breakup of nuclear material caused by collisions with high-energy photons.

photon the quantum, or unit, of light energy.

physical change any transformation that results in changes in a substance's physical state, color, temperature, dimensions, or other physical properties; the chemical identity of the substance remains unchanged in the process.

physical state the condition of a chemical substance being either a solid, liquid, or gas.

polymer a substance having large molecules consisting of repeating units.

positron a particle that is identical to an electron except that it has positive charge.

product the compounds that are formed as the result of a chemical reaction.

proton the positively charged subatomic particle found in the nuclei of atoms.

quantum a unit of discrete energy on the scale of single atoms, molecules, or photons of light.

radioactive decay the disintegration of an atomic nucleus accompanied by the emission of a subatomic particle or gamma ray.

rare earth element the metallic elements found in the two bottom rows of the periodic table; the chemistry of their ions is determined by electronic configurations with partially filled f subshells. (See also actinides; lanthanides.)

reactant the chemical species present at the beginning of a chemical reaction that rearrange atoms to form new species.

reducing agent a chemical reagent that causes an element in another reagent to be reduced to a lower oxidation state.

reduction a decrease in an atom's oxidation state; accomplished by a gain of electrons or a decrease in the number of chemical bonds to atoms of other elements. (See also oxidation state.)

rodenticide a poisonous substance intended to kill mice and rats.

semimetal another name for metalloid.

shell all of the orbitals that have the same value of the principal energy level, n.

silicate a material that consists of silicon and oxygen atoms bonded to each other.

silicone a rubberlike material that is an organic polymer of silicon.

smelter an apparatus used to produce a metal from its ore.

spallation the breaking apart of a substance.

spectrum the range of electromagnetic radiation arranged in order of wavelengths or frequencies; for example, the visible spectrum, in order of increasing frequency, exists in the order: red, orange, yellow, green, blue, indigo, and violet.

spontaneous fission the fission of a nucleus without the event having been initiated by human activity.

stardust the small component of interstellar dust that thermally condenses from hot stellar vapor as it is cooled by expansion.

subatomic particle the particles that make up an atom.

sublimation the change of physical state in which a substance goes directly from the solid to the gas without passing through a liquid state.

sublime See sublimation.

subshell all of the orbitals of a principal shell that lie at the same energy level.

supernova the explosive death of stars that have masses greater than about 10 times the mass of the Sun. All elements heavier than iron are made in supernova explosions.

tau particle an elementary particle identical to the muon, except that it has 17 times the mass. (See also muon.)

thermal conductivity a measure of the ability of a substance to conduct heat.

thermocouple a temperature-measuring device that employs the voltage difference between two different metals.

tonne also called a metric ton; consists of 1,000 kilograms.

transition metal any of the metallic elements found in the 10 middle columns of the periodic table to the right of the alkaline earth metals; the chemistry of their ions largely is determined by electronic configurations with partially filled d subshells.

transmutation the conversion by way of a nuclear reaction of one element into another element; in transmutation, the atomic number of the element must change.

transuranium element any element in the periodic table with an atomic number greater than 92.

triad any group of three elements that exhibit very similar chemical and physical properties.

ultraviolet "beyond the violet"—the region of the electromagnetic spectrum that begins where violet light leaves off and is higher in energy and frequency than violet light.

volatile description of a liquid that evaporates readily at room temperature.

Further Resources

The following sources offer readings related to individual post-transition metal and metalloid elements.

ALUMINUM

Books and Articles

Choate, William T., and John A. S. Green. "U.S. Energy Requirements for Aluminum Production: Historical Perspective, Theoretical Limits and New Opportunities." Report for the U.S. Dept. of Energy, February 2003. This report gives details about energy use in every aspect of aluminum production.

Zinner, E., and C. Göpel. "Aluminum-26 in H4 Chondrites: Implications for Its Production and Its Usefulness as a Fine-Scale Chronometer for Early Solar System Events." *Meteoritic & Planetary Science* 37 (2002): 1,001–1,013. An accessible article that explains how aluminum 26 can help date events that occurred in the solar system.

Internet Resources

Lenntech. "Aluminum—Al." Available online. URL: www.lenntech.com/Periodic-chart-elements/Al-en.htm. Accessed December 4, 2009. This article describes the applications, health effects, and environmental effects of aluminum.

Plunkert, Patricia A. "Aluminum Recycling in the United States in 2000." Available online. URL: pubs.usgs.gov/circ/c1196w/c1196w.pdf. Accessed December 4, 2009. As one of a series of reports on metals recycling, report 1196-W from the U.S. Geological Survey Circular discusses the flow of aluminum from production through its uses, with particular emphasis on the recycling of industrial scrap (new scrap) and used products (old scrap) in 2000.

GALLIUM

Books and Articles

Hebner, Robert E., Jr., J. Daniel Jones, and Kaare J. Nygaard. "GaAs Laser Experiments for the Undergraduate Laboratory." *American*

Journal of Physics 41, no. 2 (February 1973): 217–221. A set of laboratory experiments designed to investigate the properties of gallium arsenide lasers is reported.

Leutwyler, Kristin. "Foolproof Quantum Cryptography." *Scientific American,* 22 December 2000. Article describes how gallium arsenide electronics can help keep intruders from retrieving computer information.

Schwarzschild, Bertram. "Gallex Data Can't Quite Lay the Solar Neutrino Problem to Rest." *Physics Today* 45, no. 8 (August 1992). Summarizes the findings from the Gallex experiment.

Internet Resources

Lenntech. "Gallium—Ga." Available online. URL: www.lenntech.com/Periodic-chart-elements/Ga-en.htm. Accessed April 11, 2009. This article describes the applications, health effects, and environmental effects of gallium.

Photovoltaic Systems Research & Development, Sandia National Laboratories. "Gallium Arsenide Solar Cells." Available online. URL: photovoltaics.sandia.gov/docs/PVFSCGallium_Arsenide_Solar_Cells.htm. Accessed December 4, 2009. An article that describes how GaAs is used in solar cells.

United States Geological Survey. "Gallium." Available online. URL: minerals.usgs.gov/minerals/pubs/commodity/gallium/mcs-2009-galli.pdf. Accessed December 4, 2009. This is an article on the latest USGS information on gallium as a commodity in the United States, including domestic production and use, prices, events, trends, and issues.

INDIUM AND THALLIUM

Books and Articles

Levi, Barbara G. "New Solar Energy Cells May Prove Economical." *Physics Today* 28, no. 5 (May 1975). This article describes two new solar cells that could turn out to be many times cheaper than silicon cells.

Schwarz-Schampera, Ulrich, and Peter M Herzig. *Indium.* Heidelberg, Germany: Springer Verlag, 2002. A comprehensive yet accessible account of indium's properties, production, mining, and uses.

Internet Resources

Gorder, Pam Frost. "Material May Help Autos Turn Heat into Electricity." July 31, 2008. Available online. URL: www.mecheng.osu.edu/news/material-may-help-autos-turn-heat-electricity. Accessed December 4, 2009. Article describes how thallium-doped lead telluride may be used to turn heat into electricity.

Lenntech. "Indium—In." Available online. URL: www.lenntech.com/Periodic-chart-elements/In-en.htm. Accessed December 4, 2009. This site describes the applications, health effects, and environmental effects of indium.

Lenntech. "Thallium—Tl." Available online. URL: www.lenntech.com/Periodic-chart-elements/Tl-en.htm. Accessed December 4, 2009. This site describes the applications, health effects, and environmental effects of thallium.

TIN

Books and Articles

Drews, Robert. *The End of the Bronze Age: Changes in Warfare and the Catastrophe ca. 1200 B.C.* Princeton, N.J.: Princeton University Press, 1993. This book describes the cultural upheavals that accompanied the shift from the Bronze Age to the Iron Age.

Gielen, Marcel, Alwyn G. Davies, Keith Pannell, and Edward Tiekink, eds. *Tin Chemistry: Fundamentals, Frontiers, and Applications.* Hoboken, N.J.: Wiley, 2008. First authoritative one-volume survey of all aspects of modern tin chemistry, appropriate for the specialist and nonspecialist alike.

Harms, William. "Bronze Age Source of Tin Discovered." *The University of Chicago Chronicle* 13, no. 9 (January 6, 1994). This article chronicles a discovery that tin mining was a well-developed industry in the Central Taurus Mountains in Turkey as long ago as 2870 B.C.E.

Leutwyler, Kristin. "Tin Foils Melting Ideas." *Scientific American,* 22 September 2000. This article describes what scientists have learned about the strange nature of tin nanoclusters.

Penhallurick, R.D. *Tin in Antiquity: Its Mining and Trade Throughout the Ancient World with Particular Reference to Cornwall.* London, U.K.: Maney Publishing, 2008. This is the first comprehensive history of the early metallurgy of tin, a mine of information on this rare, highly prized metal so vital to the developing civilization of the Bronze Age.

Internet Resources

Lenntech. "Tin—Sn." Available online. URL: www.lenntech.com/Periodic-chart-elements/Sn-en.htm. Accessed December 4, 2009. This site describes the applications, health effects, and environmental effects of tin.

New Scientist. "Science: Tin Makes Its Interstellar Debut." Available online. URL: www.newscientist.com/article/mg13818673.000-science-tin-makes-its-interstellar-debut.html. Accessed December 4, 2009. Article discusses the method for detecting interstellar tin.

LEAD AND BISMUTH

Books and Articles

Kessel, Irene, and John T. O'Connor. *Getting the Lead Out: The Complete Resource for Preventing and Coping with Lead Poisoning.* Cambridge, Mass.: Perseus, 2001. This book covers common household sources of lead, health effects, and federal laws and regulations regarding lead poisoning.

Warren, Christian. *Brush with Death: A Social History of Lead Poisoning.* Baltimore, Md.: The Johns Hopkins University Press, 2000. This book argues that lead poisoning in all its guises was mostly silenced by design over the past century. The author covers the history of lead use and its effects in great detail.

Internet Resources

Lenntech. "Bismuth—Bi." Available online. URL: www.lenntech.com/Periodic-chart-elements/Bi-en.htm. Accessed December 4, 2009. This site describes the applications, health effects, and environmental effects of bismuth.

Lenntech. "Lead—Pb." Available online. URL: http://www.lenntech.com/Periodic-chart-elements/Pb-en.htm. Accessed December 4, 2009. This site describes the applications, health effects, and environmental effects of lead.

BORON

Books and Articles

Gupta, Umesh C. *Boron and Its Role in Crop Production.* Boca Raton, Fla.: CRC Press, 1993. This book offers in-depth coverage of the chemistry of boron, the extraction of boron from various soils, methods for determining boron in soil and plants, and the role of boron in the physiology of plants and seed production. It also examines the technology and application of boron fertilizers for crops, the response to boron of various crops, boron deficiency and toxicity in plants, and boron distribution among plant parts.

Miller, Johanna. "New Sheet Structures May Be the Basis for Boron Nanotubes." *Physics Today* 60 (November 20, 2007). Article describes the advantages of using boron in nanotube structures.

Internet Resources

Lenntech. "Boron—B." Available online. URL: www.lenntech.com/Periodic-chart-elements/B-en.htm. Accessed December 4, 2009. This site describes the applications, health effects, and environmental effects of boron.

National Boron Research Institute. "History of Boron Mining." Available online. URL: www.boren.gov.tr/en/tarihce.htm. Accessed online December 4, 2009. A timeline of boron mining from the mid-1700s to 2004.

SILICON AND GERMANIUM

Books and Articles

Jaroniec, Mietek. "Silicon Beyond the Valley." *Nature Chemistry* 1, no. 2 (May 2009): 166. In this article, the author describes how silicon, either as the pure metal or bonded to other materials, continues to play a valuable role in modern technology.

Schwarzschild, Bertram M. "New Uses for Amorphous Silicon." *Physics Today* 32, no. 12 (December 1979). Article discusses early investigations into the use of amorphous silicon for solar panels.

Seitz, Frederick. "Research on Silicon and Germanium in World War II." *Physics Today* 48 (January 1995). This article discusses how the work on the electrical properties of silicon and germanium during World War II by a relatively large and well-funded group of investigators completely transformed attitudes toward the physical properties of the pure crystalline forms of the semiconductors.

Internet Resources

Lenntech. "Germanium—Ge." Available online. URL: www.lenntech.com/Periodic-chart-elements/Ge-en.htm. Accessed December 4, 2009. This site describes the applications, health effects, and environmental effects of germanium.

Lenntech. "Silicon—Si." Available online. URL: www.lenntech.com/Periodic-chart-elements/Si-en.htm. Accessed December 4, 2009. This site describes the applications, health effects, and environmental effects of silicon.

ARSENIC AND ANTIMONY

Books and Articles

Meharg, Andrew. *Venomous Earth: How Arsenic Caused the World's Worst Mass Poisoning.* New York, N.Y.: Macmillan, 2005. An account of the disastrous accidental poisoning of tens of thousands of Bangladeshi villagers.

Naidu, R., Euan Smith, Gary Owens, Prosun Bhattacharya, and Peter Nadebaum, eds. *Managing Arsenic in the Environment: From Soil to Human Health.* Enfield, N.H.: Science Publishers, 2006. This book brings together the current knowledge on arsenic contamination worldwide, reviewing the field, highlighting common themes, and pointing to key areas needing future research.

Internet Resources

Lenntech. "Antimony—Sb." Available online. URL: www.lenntech.com/Periodic-chart-elements/Sb-en.htm. Accessed December 4,

2009. This site describes the properties, health effects, and environmental effects of antimony.

Lenntech. "Arsenic—As." Available online. URL: www.lenntech.com/Periodic-chart-elements/As-en.htm. Accessed December 4, 2009. This site describes the properties, health effects, and environmental effects of arsenic.

TELLURIUM AND POLONIUM

Books and Articles

Biello, David. "Solar Power Lightens Up with Thin-Film Technology." *Scientific American,* 25 April 2008. This article describes how thin-film photovoltaic cells may evolve to cheap, durable, efficient devices to generate a significant amount of electricity from the Sun.

Internet Resources

Lenntech. "Polonium—Po." Available online. URL: www.lenntech.com/Periodic-chart-elements/Po-en.htm. Accessed December 4, 2009. This site describes the properties, health effects, and environmental effects of polonium.

Lenntech. "Tellurium—Te." Available online. URL: www.lenntech.com/Periodic-chart-elements/Te-en.htm. Accessed December 4, 2009. This site describes the properties, health effects, and environmental effects of tellurium.

General Resources

The following sources discuss general information on the periodic table of the elements.

Books and Articles

Ball, Philip. *The Elements: A Very Short Introduction.* New York, N.Y.: Oxford University Press, 2004. This book contains useful information about the elements in general.

Chemical and Engineering News 86, no. 27 (July 2, 2008). A production index published annually showing the quantities of various chemicals that are manufactured in the United States and other countries.

Considine, Douglas M., ed. *Van Nostrand's Encyclopedia of Chemistry,* 5th ed. New York, N.Y.: John Wiley and Sons, 2005. In addition to its coverage of traditional topics in chemistry, the encyclopedia has articles on nanotechnology, fuel cell technology, green chemistry, forensic chemistry, materials chemistry, and other areas of chemistry important to science and technology.

Cotton, F. Albert, Geoffrey Wilkinson, and Paul L. Gaus. *Basic Inorganic Chemistry,* 3rd ed. New York, N.Y.: John Wiley and Sons, 1995. Written for a beginning course in inorganic chemistry, this book presents information about individual elements.

Cox, P. A. *The Elements on Earth: Inorganic Chemistry in the Environment.* New York, N.Y.: Oxford University Press, 1995. There are two parts to this book. The first part describes Earth and its geology and how elements and compounds are found in the environment. Also, it describes how elements are extracted from the environment. The second part describes the sources and properties of the individual elements.

Daintith, John, ed. *The Facts On File Dictionary of Chemistry,* 4th ed. New York, N.Y.: Facts On File, 2005. Definitions of many of the technical terms used by chemists.

Downs, A. J., ed. *Chemistry of Aluminium, Gallium, Indium and Thallium.* New York, N.Y.: Springer, 1993. A detailed, wide-ranging, authoritative and up-to-date review of the chemistry of aluminum, gallium, indium, and thallium. Coverage is of the chemistry and commercial aspects of the elements themselves; emphasis is on the design and synthesis of materials, their properties, and their applications.

Emsley, John. *Nature's Building Blocks: An A–Z Guide to the Elements.* New York, N.Y.: Oxford University Press, 2001. Proceeding through the periodic table in alphabetical order of the elements, Emsley describes each element's important properties, biological and medical roles, and importance in history and the economy.

———. *The Elements.* New York, N.Y.: Oxford University Press, 1989. In this book, Emsley provides a quick reference guide to the chemical, physical, nuclear, and electron shell properties of each of the elements.

Foundations of Chemistry 12, no. 1 (April 10, 2010). This special issue of the journal focuses on the periodic table, featuring some obscure history, possible new arrangement, and the role of chemical triads.

Greenberg, Arthur. *Chemistry: Decade by Decade.* New York, N.Y.: Facts On File, 2007. An excellent book that highlights by decade the important events that occurred in chemistry during the 20th century.

Greenwood, N. N., and A. Earnshaw. *Chemistry of the Elements.* Oxford, U.K.: Pergamon Press, 1984. This book is a comprehensive treatment of the chemistry of the elements.

Hall, Nina, ed. *The New Chemistry.* Cambridge, U.K.: Cambridge University Press, 2000. This book contains chapters devoted to the properties of metals and electrochemical energy conversion.

Hampel, Clifford A., ed. *The Encyclopedia of the Chemical Elements.* New York, N.Y.: Reinhold Book Corp., 1968. In addition to articles about individual elements, this book also has articles about general topics in chemistry. Numerous authors contributed to this book, all of whom were experts in their respective fields.

Heiserman, David L. *Exploring Chemical Elements and Their Compounds.* Blue Ridge Summit, Penn.: Tab Books, 1992. This book is described by its author as "a guided tour of the periodic table for ages 12 and up," and is written at a level that is very readable for pre-college students.

Henderson, William. *Main Group Chemistry*. Cambridge, U.K.: The Royal Society of Chemistry, 2002. This book is a summary of inorganic chemistry in which the elements are grouped by families.

Jolly, William L. *The Chemistry of the Non-Metals*. Englewood Cliffs, N.J.: Prentice-Hall, 1966. This book is an introduction to the chemistry of the nonmetals, including the elements covered in this book.

King, R. Bruce. *Inorganic Chemistry of Main Group Elements*. New York, N.Y.: Wiley-VCH, 1995. This book describes the chemistry of the elements in the "s" and "p" blocks.

Krebs, Robert E. *The History and Use of Our Earth's Chemical Elements: A Reference Guide*, 2nd ed. Westport, Conn.: Greenwood Press, 2006. Following brief introductions to the history of chemistry and atomic structure, Krebs proceeds to discuss the chemical and physical properties of the elements group (column) by group. In addition, he describes the history of each element and current uses.

Lide, David R., ed. *CRC Handbook of Chemistry and Physics*, 89th ed. Boca Raton, Fla.: CRC Press, 2008. The *CRC Handbook* has been the most authoritative, up-to-date source of scientific data for almost nine decades.

Mendeleev, Dmitri Ivanovich. *Mendeleev on the Periodic Law: Selected Writings, 1869–1905*. Mineola, N.Y.: Dover, 2005. This English translation of 13 of Mendeleev's historic articles is the first easily accessible source of his major writings.

Minkle, J. R. "Element 118 Discovered Again—for the First Time." *Scientific American*, 17 October 2006. This article describes how scientists in California and Russia fabricated element 118.

Norman, Nicolas C. *Periodicity and the p-Block Elements*. New York, N.Y.: Oxford University Press, 1994. This book describes group properties of post-transition metals, metalloids, and nonmetals.

Parker, Sybil P., ed. *McGraw-Hill Encyclopedia of Chemistry*, 2nd ed. New York, N.Y.: McGraw Hill, 1993. This book presents a comprehensive treatment of the chemical elements and related topics in chemistry, including expert-authored coverage of analytical chemistry, biochemistry, inorganic chemistry, physical chemistry, and polymer chemistry.

Rouvray, Dennis H., and R. Bruce King, eds. *The Periodic Table: Into the 21st Century.* Baldock, Hertfordshire, U.K.: Research Studies Press Ltd., 2004. A presentation of what is happening currently in the world of chemistry.

Stwertka, Albert. *A Guide to the Elements,* 2nd ed. New York, N.Y.: Oxford University Press, 2002. This book explains some of the basic concepts of chemistry and traces the history and development of the periodic table of the elements in clear, nontechnical language.

Van Nostrand's Encyclopedia of Chemistry, 5th ed. Edited by Glenn D. Considine. Hoboken, N.J.: Wiley and Sons, 2005. This is a compendium of modern chemistry covering topics from green chemistry to nanotechnology.

Winter, Mark J., and John E. Andrew. *Foundations of Inorganic Chemistry.* New York, N.Y.: Oxford University Press, 2000. This book presents an elementary introduction to atomic structure, the periodic table, chemical bonding, oxidation and reduction, and the chemistry of the elements in the s, p, and d blocks; in addition, there is a separate chapter devoted just to the chemical and physical properties of hydrogen.

Internet Resources

About.com: Chemistry. Available online. URL: chemistry.about.com/od/chemistryfaqs/f/element.htm. Accessed December 4, 2009. Information about the periodic table, the elements, and chemistry in general from the New York Times Company.

American Chemical Society. Available online. URL: portal.acs.org/portal/acs/corg/content. Accessed December 4, 2009. Many educational resources are available here.

Center for Science and Engineering Education, Lawrence Berkeley Laboratory, Berkeley, California. Available online. URL: www.lbl.

gov/Education. Accessed December 4, 2009. Contains educational resources in biology, chemistry, physics, and astronomy.

Chemical Education Digital Library. Available online. URL: www.chemeddl.org/index.html. Accessed December 4, 2009. Digital content intended for chemical science education. Chemical Elements. Available online. URL: www.chemistryexplained.com/elements. Accessed December 4, 2009. Information about each of the chemical elements.

Chemical Elements.com. Available online. URL: www.chemicalelements.com. Accessed December 4, 2009. A private site that originated with a school science fair project.

Chemicool. Available online. URL: www.chemicool.com. Accessed December 4, 2009. Information about the periodic table and the chemical elements, created by David D. Hsu of the Massachusetts Institute of Technology.

Department of Chemistry, University of Nottingham, United Kingdom. Available online. URL: www.periodicvideos.com. Accessed December 4, 2009. Short videos on all of the elements can be viewed.

Journal of Chemical Education, Division of Chemical Education, American Chemical Society. Available online. URL: jchemed.chem.wisc.edu/HS/index.html. Accessed December 4, 2009. The Web site for the premier online journal in chemical education.

Lenntech. Available online. URL: www.lenntech.com/Periodic-chart.htm. Accessed December 4, 2009. Contains an interactive, printable version of the periodic table.

Los Alamos National Laboratory. Available online. URL: periodic.lanl.gov/default.htm. Accessed December 4, 2009. A resource on the periodic table for elementary, middle school, and high school students.

Mineral Information Institute. Available online. URL: www.mii.org. Accessed December 4, 2009. A large amount of information for teachers and students about rocks and minerals and the mining industry.

National Nuclear Data Center, Brookhaven National Laboratory. Available online. URL: www.nndc.bnl.gov/content/HistoryOf Elements.html. Accessed December 4, 2009. A worldwide resource for nuclear data.

The Periodic Table of Comic Books, Department of Chemistry, University of Kentucky. Available online. URL: www.uky.edu/Projects/ Chemcomics. Accessed December 8, 2009. A fun, interactive version of the periodic table.

The Royal Society of Chemistry. URL: http://www.rsc.org/chemsoc/. Accessed December 4, 2009. This Web site contains information about many aspects of the periodic table of the elements.

Schmidel & Wojcik: Web Weavers. Available online. URL: quizhub. com/quiz/f-elements.cfm. Accessed December 4, 2009. A K–12 interactive learning center that features educational quiz games for English language arts, mathematics, geography, history, earth science, biology, chemistry, and physics.

United States Geological Survey. Available online. URL: minerals.usgs. gov. Accessed December 4, 2009. The official Web site of the Mineral Resources Program.

University of Nottingham. Available online. URL: www.periodicvideos. com/. Accessed December 4, 2009. Short videos on all of the elements can be viewed here via the department of chemistry.

Web Elements, The University of Sheffield, United Kingdom. Available online. URL: www.webelements.com/index.html. Accessed December 4, 2009. A vast amount of information about the chemical elements.

Wolfram Science. Available online. URL: demonstrations.wolfram. com/PropertiesOfChemicalElements. Accessed December 4, 2009. Information about the chemical elements from the Wolfram Demonstration Project.

Periodicals

Discover
Published by Buena Vista Magazines
114 Fifth Avenue
New York, NY 10011

Telephone: (212) 633-4400
www.discover.com
A popular monthly magazine containing easy to understand articles on a variety of scientific topics.

Nature
The Macmillan Building
4 Crinan Street
London N1 9XW
Telephone: +44 (0)20 7833 4000
www.nature.com/nature
A prestigious primary source of scientific literature.

Science
Published by the American Association for the Advancement of Science
1200 New York Avenue, NW
Washington, DC 20005
Tel: (202) 326-6417
www.sciencemag.org
One of the most highly regarded primary sources for scientific literature.

Scientific American
415 Madison Avenue
New York, NY 10017
Telephone: (212) 754-0550
www.sciam.com
A popular monthly magazine that publishes articles on a broad range of subjects and current issues in science and technology.

Index

Note: *Italic* page numbers refer to illustrations.

F

G

Q

R

U

V

W

Z